Current Topics in Microbiology and Immunology

179

Herpes Simplex Virus

Pathogenesis, Immunobiology and Control

Edited by B. T. Rouse

With 9 Figures

Springer-Verlag
Berlin Heidelberg NewYork
London Paris Tokyo
Hong Kong Barcelona
Budapest

Barry T. Rouse, BVSc, MSc, PhD

Department of Microbiology
College of Veterinary Medicine
University of Tennessee
Knoxville, TN 37996-0845
USA

Cover Illustration: Background: Cryo electron micrograph of purified B capsids of HSV-1 protein. Inset: Three-dimensional computer reconstruction of the Mab-labelled capsid, cut away to reveal a central cross section of the capsid shell and the inner capsid surface. The Fab fragments bound to the protruding tips of hexons are color-coded red. Preparation of capsids and antibodies: W.W. Newcomb and J.C. Brown, University of Virginia. Structural analysis and computer graphics: F.P. Booy, J.F. Conway, A.C. Steven and B.L. Trus, National Institutes of Health.

ISBN-13: 978-3-642-77249-8 e-ISBN-13: 978-3-642-77247-4
DOI: 10.1007/978-3-642-77247-4

Library of Congress Catalog Card Number 15-12910
Softcover reprint of the hardcover 1st edition 1992

Typesetting: Thomson Press (India) Ltd, New Delhi; Offsetprinting: Saladruck, Berlin; Bookbinding: Lüderitz & Bauer, Berlin
23/3020-5 4 3 2 1 0 – Printed on acid free paper.

Preface

Although upstaged by the tragic appearance of the human immunodeficiency virus, herpes simplex viruses (HSV) types 1 and 2 continue to be major human pathogens against which we lack acceptable vaccines or other means of immunological control. The virus is large and complex, coding for 70 or more proteins. Although many mysteries remain to be unraveled, our knowledge base regarding genomic organization, gene expression and regulation, pathogenesis, and immune recognition of component parts is quite considerable. Indeed, meetings devoted entirely to herpesviruses are conspicuous by their frequency and excellent, yet sometimes exclusive, attendance. The purpose of this volume is to compile in a single book a series of reviews by leading investigators that deal with various aspects of virus-host interactions and which hopefully will provide clues as to how to best manage HSV from an immunobiological perspective. Ultimately, one anticipates that a full understanding of virus-host interaction will lead to strategies useful for the prevention and control of HSV. The state of current progress with conventional vaccines is presented, as is a chapter on intracellular immunization. This latter novel approach to virus infections comes at approximately the bicentenary of Jenner's introduction of a successful conventional immunization strategy.

Since effective antiviral vaccines, immunomodulators, and intracellular immunogens demand a detailed understanding of the molecular biology of HSV, the volume begins with a chapter by HAY and RUYECHAN which briefly presents an overview of HSV gene expression and mentions the protein products that could be targeted for conventional immunological or intracellular attack. Following this are two chapters on pathogenesis. The first, by STANBERRY, deals with the many disease situations which involve HSV and the experimental animal models used to investigate them. The second, by SIMMONS, TSCHKARKE, and SPECK, focuses on a major target organ of HSV infection, the nervous system. The review presents a detailed discussion of how the immune system interacts with HSV in the nervous system. The next three chapters are concerned with immunobiology and discuss the role of various host defense mechanisms in preventing and controlling infection. Thus, SCHMID and ROUSE deal with T cell immunity, emphasizing the relevance of cytotoxic T lymphocytes. KOHL discusses antibody-mediated immunity particularly as it relates to neonatal infections. Finally, WU and MORAHAN discuss the many aspects of nonadaptive immunity which play a vital role in containing HSV infection. The subsequent chapter introduces the intriguing possibility that the virus has evolved means of avoiding or dealing with the immune response. In this review, DUBIN and coworkers discuss instances in which viral properties appear to diffuse the otherwise protective effects of the immune response. The following chapter by DOYMAZ and ROUSE points out situations in which the immune response appears to contribute to disease lesions. Such immunopathological events are particularly damaging to the eye. BURKE summarizes our state of progress with vaccine development and considers both vaccines to prevent infection and those claimed to be therapeutically effective against HSV. The volume ends with a chapter by WONG and CHATTERJEE on the exciting new topic of intracellular immunization, which may represent the future of antiviral control. Their chapter is a general one, but it does focus on their own recent studies on HSV.

BARRY T. ROUSE

Contents

Contributors

(Their addresses can be found at the beginning of their respective chapters)

Regulation of Herpes Simplex Virus Type 1 Gene Expression

J. HAY and W. T. RUYECHAN

1 Introduction

The *Herpesviridae* represent a large and diverse family of vertebrate pathogens and at least seven members of the group infect humans (ROIXMAN 1990a). In this brief review, we shall discuss gene expression in herpes simplex virus (HSV) and point out those gene products which act as targets for the immune response.

All herpes virions consist of an outer lipid membrane surrounding an amorphous, proteinaceous structure known as the tegument, which in turn surrounds an icosahedral nucleocapsid (HAY et al. 1987). The lipid envelop contains about eight virus-encoded glycoproteins, which probably act as the principal components that stimulate humoral immune responses. The tegument is the least characterized portion of the virion structure, and while it has classically been characterized as being composed of protein, it appears contiguous with the lipid membrane of the virus and thus could potentially contain phospholipid molecules as well. As will be discussed below, the tegument contains at least two proteins involved in the transactivation of gene expression and one involved in the shutoff of host macromolecular synsthesis. The nucleocapsids of all

Department of Microbiology, State University of New York at Buffalo, Buffalo, New York 14214, USA

Current Topics in Microbiology and Immunology, Vol. 179

herpes virions are approximately 100 nm in diameter and are made up of 162 hexameric and pentameric capsomers. The major component is termed the major capsid protein and the molecular weights of these proteins are approximately 150–200 kDa. Five to six additional proteins are present in the capsid which seen to be required for the generation of capsid structures capable of packaging newly synthesized viral DNA. The core of mature herpes virions contains the viral DNA, which appears to be packaged as a tightly wrapped structure. While earlier studies indicated the presence of a less dense, possibly proteinaceous core around which a toroid of DNA was wound, more recent work (BOOY et al. 1991) suggests that DNA alone forms the core of the virion and is probably organized as a "ball of wool," as are the T-even bacteriophage genomes.

2 HSV Type 1 Regulatory Cascade

The HSV-1 lytic cycle is approximately 18 h in length. Infection begins with the fusion of the virion envelope with the target cell plasma membrane. The capsid and at least a portion of the tegument are then transported to the nuclear envelope where the viral DNA and one or more of the tegument proteins enter the nucleus. A minimum of five HSV-1 proteins are then synthesized in a sequential coordinately regulated scheme. Five proteins known as α or immediate early (IE) proteins, are expressed without any prior synthesis of viral protein. Peak synthesis occurs 2–4 h postinfection, but some α proteins continue to be produced throughout infection. At least three of these proteins go on to transactivate and/or transrepress the expression of the remaining four classes of virus proteins. Immediate early gene products may act as targets for cytotoxic T lymphocytes (MARTIN et al. 1988) and perhaps also for natural killer cell lysis (FITZGERALD-BOCARSLY et al. 1991).

One major kinetic class of viral proteins is comprised of two groups of β or early (E) polypeptides, $\beta 1$ and $\beta 2$. The β polypeptides are almost exclusively involved in nucleic acid metabolism. While, as a whole, they are synthesized at peak rates approximately 5–7 h postinfection, the $\beta 1$ class is synthesized earlier, its peak synthesis overlapping slightly with that of α class. The $\beta 1$ proteins, however, require the presence of at least one of the α proteins ($\alpha 4$ or IE175) for their synthesis. The $\beta 2$ class is synthesized at peak rates 6–7 h postinfection. Replication of the 152 kb viral DNA begins at the temporal interface of α and β protein synthesis and continues for approximately another 12 h. The peak rates of viral DNA synthesis appear to occur relatively early in infection, although, as pointed out by ROIZMAN (1990a), this is difficult to assess with standard radioisotopic labeling of newly synthesized viral DNA due to rapid saturation of nucleotide pools in infected cells (ROIZMAN and ROANE 1964).

Another kinetic class of proteins are the γ or late (L) class polypeptides, $\gamma 1$ and $\gamma 2$. The γ proteins are primarily structural polypeptides and include the viral glycoproteins, capsid, and some tegument components. Differentiation of the $\gamma 1$ and $\gamma 2$ classes is based upon the dependence of their expression on DNA synthesis: $\gamma 1$ proteins can be synthesized in the absence of viral DNA replication although at considerably reduced levels compared to those produced under conditions of productive infection. The $\gamma 2$ class proteins, in contrast, absolutely require viral DNA replication for their synthesis.

Following transcription of the γ genes, empty capsids are assembled and newly synthesized DNA, which is believed to be present as large head-to-tail concatamers, is cleaved to genome size length and packaged into the nucleocapsids by an as yet unknown mechanism. The full capsids are then enveloped via attachment to the inner lamella of the nuclear membrane. The specific point at which the tegument proteins become associated with the nucleocapsid during this process is not presently known. The virions then move to the cell surface for release via a mechanism that may involve secretion through the Golgi apparatus (JOHNSON and SPEAR 1983).

Based on all available data, the RNA polymerase which transcribes herpesvirus DNAs is the cellular RNA polymerase II. The viral mRNAs are, with a few exceptions, unspliced transcripts. The posttranscriptional modifications of HSV-1 mRNAs are typical of those for eukaryotic messages and include capping, methylation, and polyadenylation (STRINGER et al. 1979). Similarly, virus-encoded proteins undergo posttranscriptional modification following synthesis on both free and bound ribosomes. Posttranscriptional modifications include cleavage, phosphorylation, sulfation, O- and N-linked glycosylation, and myristylation (ACKERMAN et al. 1984; HOPE and MARSDEN 1983). The majority of α, β, and γ viral proteins thus far characterized appear to be modified in some fashion after synthesis.

We will now turn to a more detailed analysis of the expression and functions of specific gene products in each group. It should be borne in mind from the outset of this part of the discussion that the majority of the proteins discussed are required for the synthesis of HSV-1 in tissue culture. These gene products represent 50%–70% of the over 70 open reading frames (ORFs) identified in the sequence of the HSV-1 genome. The remaining proteins and predicted proteins have been termed supplementary essential genes by ROIZMAN (1990b). These gene products, while dispensible for growth in tissue culture, were either shown to have or are believed to have required functions in the human or animal host. Hence, they should be considered to be at least as important as those genes essential for growth in tissue culture and clearly of potential value in the context of targets for immunization.

3 Expression and Function of the α (IE) Proteins

The viral DNA which is transported through the nuclear pores during HSV-1 infection is a linear duplex molecule made up of two segments of unique sequences, 67 and 8.9 kb in length, known as the unique long (U_L) and unique short (U_S) regions, respectively. They are bounded by pairs of inverted repeats, 6.3 kb in length for U_L and 4.3 kb for U_S. Upon entering the nucleus, the viral DNA is circularized via the action of cellular and (possibly) viral tegument/capsid proteins (POFFENBERGER and ROIZMAN 1985) and the α genes are transcribed de novo. These include *α4*, *α0*, *α27*, *α22*, and *α47*. Alternate designations for these genes are *IE175*, *IE110*, *IE63*, *IE68*, and *IE12*, respectively. *α4* maps in the inverted repeat bounding the U_S region and *α0* maps in the inverted repeat bounding the U_L region; thus, these genes are diploid in the HSV genome. *α27* maps in the U_L region and *α22* and *α47* map in the U_S region.

The *cis*-acting elements required for efficient expression of the α genes, located upstream of the transcription start sites, are the most complex and extensive of the HSV promoters. All include elements similar to those found in many eukaryotic promoters; an enhancer element unique to the α gene promoters is located upstream (some 300–500 bp) from the transcription start site. This element conforms to the consensus sequence TAATGARAT (GAFFNEY et al. 1985). The enhancer sequence TAATGARAT acts as a binding site for the cellular transcription factor known as OCT-1 and is necessary and sufficient for the transactivation of the α genes by a virion tegument protein known as α-transinducing factor (α-TIF). The currently available data indicate that α-TIF alone is incapable of binding to DNA. Recent experiments (reviewed by ROIZMAN and SEARS 1990) indicate that a GA-rich element of the enhancer is required for the full activity of α-TIF. The active portion of the α-TIF molecule appears to be the acidic 81 amino acid COOH terminal of the protein.

The α gene product most intensively studied is the *α4* protein. It has a predicted molecular mass of 140 kDa and an apparent molecular mass of 170–175 kDa, with three major species (*α4*a, *α4*b, and *α4*c). The high molecular mass and presence of multiple forms are due in part to phosphorylation (and probably other posttranslational modifications) of the protein, which occurs upon its translocation to the nucleus. *α4* is essential for viral replication and is required for both the activation and transcription of β and γ genes and for autoregulation (repression) of its own synthesis. Extensive biochemical studies have show that the active form of the protein is a homodimer which retains an intrinsic affinity for DNA. Recently, it has been shown that both *α4* and its Varicella Zoster Virus (VZV) equivalent (*IE62*) are present in virions (YOA and COURTNEY 1989; KINCHINGTON et al. 1992). This suggests that *α4* may be required immediately for an efficient beginning to the infectious cycle.

α4 has been shown to bind to the sequence ATCGTCNNNNNCGRC. Such sequences are present in several HSV promoters, such as the promoter/regulatory sequences of the *α4* gene itself and those of *α0*. *α4* protein is required for the efficient expression of some β and γ genes which do not contain the consensus sequence in their upstream regions. Thus, additional binding sites or an alternate mechanism of *α4* action must be proposed. Also, there is no apparent correlation between the consensus binding site and sequences required for transactivation. Thus, either there are alternative *α4* binding sites used in the transactivation of genes whose promoters lack the consensus sequence or there are other viral or cellular factors which mediate the transactivation. This latter model is consistent with some early data which indicated that purified *α4* required host factors for interaction with calf thymus DNA columns. *α4* appears to be a major target for cytotoxic T lymphocytes in at least one strain of mice (MARTIN et al. 1990).

The protein product of the second α gene, *α0*, has been defined as a "promiscuous" *trans*-activator based on its ability to transactivate a variety of genes in transient assays. This transactivation appears to be enhanced by the presence of *α4*. The *α0* gene is diploid; the protein has a predicted size of 78 kDa and an apparent molecular weight of 110 kDa. This difference appears to be due in part to phosphorylation. *α0* was shown capable of binding to calf thymus DNA columns when present in crude extracts and is translocated to the nucleus where it is associated with chromatin in vivo (KNIPE 1989; HAY and HAY 1980). The predicted primary structure of *α0* indicates that its NH_2-terminal is highly acidic and that the protein contains a cysteine-rich region which conforms to the "zinc finger" motif described by BERG (1990). Both of these features have been associated with a variety of other transcriptional activators.

An extensive mutational analysis by EVERETT and coworkers (1991) has shown that the two domains which appear to be crucial for the synergistic interaction of *α0* with *α4* are the potential zinc-binding domain and the COOH-terminal of the protein. Further mutational analysis, particularly the generation of virus strains which have had all or a portion of the *α0* gene deleted, indicate that this gene product is not required for lytic infection in cell culture. Despite the remarkably efficient translational activation and synergy with *α4* protein seen in transient assays, strains lacking the *α0* gene grow only slightly less well than wild-type virus when experiments are done at high multiplicity. At low multiplicity, however, *α0* appears to be required for the significant expression of β and γ genes. Thus, at low multiplicity, the amounts of *α4* and *α27* alone (see below) do not appear to be sufficient for viral growth. The mechanism behind this requirement is not as yet understood. *α0* has not been reported to act as a target for the immune response.

A third α protein, which has the ability to affect HSV transcription in *trans*, is *α27*. The *α27* protein is capable of acting either as a transrepressor or a transactivator, depending upon the temporal class of HSV genes it is acting upon. Specifically, *α27* was shown to be a repressor of α and some β genes and is required for the expression of γ genes. Deletion and temperature-sensitive mutations within the *α27* gene have shown that it is essential for viral growth. Again, the predicted size (55 kDa) of the *α27* protein is smaller than the apparent SDS PAGE-derived molecular weight of 63 kDa, probably owing to phosphorylation. Analysis of the predicted structure of the *α27* protein (SANDRI-GOLDIN 1991) reveals a hydrophilic NH_2 half and a hydrophobic COOH half. It also contains a potential "zinc finger" within the COOH-terminal residues, similar to those seen in other *trans*-activating proteins.

SANDRI-GOLDIN (1991) showed that both the transrepressor and trans-activator activities of *α27* reside in the COOH-terminal portion of the protein. The mechanisms by which *α27* acts to carry out its role as a transrepressor and the synergistic action of this protein with *α4* and *α0* are, as yet, unknown. *α27* has also been shown to play a role in stabilization and/or export of improperly processed mRNAs (SANDRI-GOLDEN 1991). This last activity has, thus, far only been demonstrated in transient expression assays and requires the presence of plasmids expressing *α4* and *α0*. The mechanisms of action at the posttranslation level are also currently unknown. Recently, *α27* was shown to act as a target for murine cytotoxic T lymphocytes (BANKS et al. 1991).

Our knowledge concerning the two remaining α proteins, 22 and 47, is considerably less detailed. Deletion analysis of *α22* indicated a potential role for this immediate early protein in defining host range. *α22*-negative mutants were shown to grow normally in simian and human cells (POST et al. 1981), but synthesis of γ genes is suppressed in certain rodent cell lines, and virus growth in these cell lines was suboptimal (SEARS et al. 1985). Thus, *α22* may play an auxiliary role in late gene expression which can be complemented by a cellular factor in primate cells. In contrast, deletion of the *α47* gene resulted in no obvious attenuation of viral growth in numerous cell lines (MAVROMARA-NAZOS et al. 1986). Thus, *α47* serves as an example of a "supplementary essential" HSV gene. We know nothing of the roles of *α47* and *α22* as targets for the immune response.

4 Expression and Function of the β (E) Genes

The β proteins characterized to date are primarily involved in viral DNA synthesis. Seven gene products (CHALLBERG 1986) and a *cis*-acting origin element, which maps in the inverted repeat bounding the U_S region, appear to be absolutely required for replication of the HSV-1 genome. Numerous additional β gene products, primarily enzymes involved in nucleotide metabolism, and a second origin, which maps in the center of the U_L region, are at least partially dispensable for this process in tissue culture, although viruses containing mutations in many of these auxiliary genes are often somewhat attenuated for growth (WELLER 1991). The actual mechanism by which HSV DNA is replicated remains obscure, although most evidence currently available favors a rolling circle model. This appears to be clearly the case during later stages of infection, but the circularization of input viral DNA raises the formal possibility that Cairn's-type bidirectional synthesis may occur during the early stages of the viral cycle.

All β genes thus far examined appear to have similar promoters. These consists of a TATA consensus sequence and distal sites made up either of SP1-like binding sites and/or CAAT box elements (SU and KNIPE 1987). They appear to be one level less in complexity than α promoters in that the TAATGARAT element is not involved. The presence of functional α proteins, however, is absolutely required.

The functions and genetics of the seven proteins necessary and sufficient for origin-dependent DNA synthesis have been the subject of several recent, extensive, and excellent reviews (e.g. OLIVO and CHALLBERG 1991). The genes encoding the replication proteins all occur in the U_L region of the viral genome. One of these, U_L9, encodes a protein which binds to and conformationally alters the origin of DNA replication. The protein also has a DNA helicase activity in the presence of ATP and the HSV major DNA-binding protein. Its mechanism of action is to render the DNA at the origin accessible to a multienzyme complex made up of the remainder of the replication proteins.

U_L30, encodes the catalytic subunit of the viral DNA polymerase. This enzyme is multifunctional and displays DNA-dependent polynucleotide synthetase, 3′–5′ exonuclease, and RNAse H activities. This polypeptide is the subject of intensive and ongoing mutational analysis and continues to be a major target of antiviral drug development. U_L42 encodes a protein which, together with the polymerase catalytic subunit, makes up the DNA polymerase holoenzyme in a one-to-one complex. The U_L42 protein binds nonspecifically to duplex DNA and, in the context of the polymerase activity, appears to anchor the holoenzyme to the template DNA (M. CHALLBERG and C. KNOPF, personal communications). The protein is produced late into infection at levels far exceeding those necessary for stoichiometric interaction with the U_L30 gene product and hence may well have additional functions. This protein is phosphorylated and is the only one of the seven required replication genes which is posttranslationally modified.

The gene products of U_L5, U_L8, and U_L52 form a three-subunit complex which exhibit helicase and primase activities. Such activities would, by analogy with prokaryotic systems, unwind the template DNA and prime DNA synthesis on the lagging strand. It is not as yet known which polypeptide chains contain the two activities. However, recent work has shown that only the U_L5 and U_L52 proteins are required for both activities. Thus, U_L8 appears to be an accessory protein with an essential, but as yet undefined, function.

The final replication protein to be considered is the product of the U_L29 gene. This is the HSV-1 major DNA-binding protein usually designated ICP8. Based on work from numerous laboratories, ICP8 was shown to be a multifunctional protein involved in both DNA replication and gene regulation. The protein exhibits a number of properties which suggest that it is in part analogous to the bacteriophage T4 gene *32* protein which is also involved in these two fundamental processes. Like this protein, ICP8 contains a "zinc finger" motif and has recently been shown to contain one mole of tightly bound zinc per mole of protein (GUPTE et al. 1991).

ICP8 binds cooperatively and preferentially to single-stranded DNA but is also capable of interaction with duplex DNA and polyribonucleotides. It is capable of limited stimulation of the viral DNA polymerase and, as mentioned above, stimulates the helicase activity of the origin-binding protein. In addition to these properties, DE BRUYN KOPS and KNIPE (1988) have shown that ICP8 a responsible for the organization of replication complexes and for their proper entry into replication compartments in the nucleus at the onset of DNA replication. Finally, a growing body of evidence indicates that ICP8 is a significant factor in the regulation of HSV gene expression. ICP8 down-regulates β protein synthesis by decreasing viral transcription and recently has been shown to be involved in the maximal synthesis of some γ proteins (GAO and KNIPE 1991). Since both of these properties imply specific binding to viral DNA, it is possible that ICP8 may possess either sequence- or structure-specific binding modes. In support of the latter hypothesis, recent data from our laboratories indicate that ICP8 can interact with and partially unwind superhelical DNA (W. RUYECHAN, unpublished observations).

Five β genes encode polypeptides that are enzymes involved in nucleotide metabolism. These include thymidine kinase (actually pyrimidine deoxynucleotide kinase; U_L23) and ribonucleotide reductase (U_L39 and U_L40). All of these proteins are dispensable, to variable extents, for growth in tissue culture. However, they are probably required to manipulate and maintain nucleotide pools during viral infection of specialized somatic cells (such as neurons) and, hence, may be indispensable for viral replication in the host. In this context, thymidine kinase-negative virus grows poorly in peripheral sites in the mouse and cannot be reactivated from the latent state.

Finally, other activities, including a topoisomerase and a dCMP deaminase, have been reported in infected cells and would probably qualify as supplementary essential genes if proven to be of viral origin. In addition, a novel DNA primase activity, which is associated with the viral DNA polymerase in partially purified extracts, has also been reported (HOLMES et al. 1988). The source of this activity and its relationship to the essential viral helicase/primase function (see above) are unknown. We know little or nothing of the role of β proteins as stimulators or targets of the immune response.

5 Expression and Function of the γ (L) Genes

As mentioned above, late HSV genes are generally divided into two classes, $\gamma 1$ and $\gamma 2$, based upon timing of their synthesis and the effect of DNA synthesis (CONLEY et al. 1981; SILVER and ROIZMAN 1985). Probably the best criterion for "lateness" regarding proteins is kinetics of synthesis. A characteristic $\gamma 1$ gene encodes the major capsid protein (VP5), while a typical $\gamma 2$ gene product is the glycoprotein gC (HOLLAND et al. 1979).

As directed for the α and β classes of HSV genes, the expression of γ genes is controlled by a series of *cis* and *trans*-acting factors, but the mechanism of their action remains to be worked out in detail. A unique feature of the control of γ gene expression is the requirement for viral DNA synthesis acting in some fashion in *cis* (MAVROMARA-NAZOS and ROIZMAN 1987). The $\alpha 4$ protein also appears to play a role in γ gene expression, since the promoter sequences for two $\gamma 2$ genes contain $\alpha 4$ binding sites (MICHAEL et al. 1988). In support of this, transient assay experiments directly demonstrate that $\alpha 4$ can activate transcription from HSVγ promoters (EVERETT 1986). In addition, studies with viral mutants show that $\alpha 22$ and $\alpha 27$ may be involved (directly or indirectly) in the positive qualitative or quantitative control of late gene expression (EVERETT 1986). The β protein ICP8 (see above) probably also functions in late gene control in that it may activate transcription (GAO and KNIPE 1991). The $\alpha 0$ protein is able to activate both $\gamma 1$ and $\gamma 2$ promoters, acting synergistically with $\alpha 4$, while under certain circumstances, $\alpha 27$ is able to repress transcription (SEKULOVITZ et al. 1988). $\alpha 47$ plays no regulatory role with HSVγ genes. This rather confusing picture may in part be due, as pointed out by EVERETT (1966) and others, to a reliance on transient assay to elucidate control of late genes; it is now clear that such assays may not necessarily reflect the in vivo situation.

Comparison of signals required for α, β or γ gene transcriptional control reveals a "nested set" situation in whch: (a) α genes require a TATA element, distal signals (such as a CAAT box or SP1-binding sites), and an upstream TAATGARAT motif, (b) β genes require only the TATA box and distal signals, and (c) γ genes need only the TATA box. Superimposed on this, of course, is the need for 5′ noncoding regions for maximum efficiency.

What distinguishes a $\gamma 1$ from a $\gamma 2$ gene is emerging from recent experiments in which chimeric promoter elements have been constructed (HOMA et al. 1988; JOHNSON and EVERETT 1986). These suggest that lateness (i.e., ability to be expressed after DNA synthesis) is a function of the TATA element sequences, although no consensus motifs for these have yet been defined. The ability to be expressed before DNA synthesis seems to reside in sequences upstream (5′) from the TATA element, but how these operate to negate DNA replication control over expression is also unclear at present.

The HSVγ proteins are involved in the structure of the virus particle (VP) and may act as major targets for the immune response. There may be as many as 35 proteins in the HSV virion, including up to 7 in mature (enveloped) nucleocapsids: VP5, VP19C, VP21, VP22, VP23, and VP24 (SCHRAG et al. 1989). VP5 is the major capsid protein and is considered to make up the bulk of both the hexameric and pentameric capsomers. The envelope of the virus contains eight viral glycoproteins, gB, gC, gD, gE, gG, gH, gI, and gJ, only three of which (gB, gD, and gH) are essential for virus multiplication in cell culture. Many if not all of them play a major part as targets for the immune response (see the chapters by SIMMONS et al., SCHMID and ROUSE, KOHL, WU and MORAHAN). Between the capsid of the virus and its envelope resides the tegument, an apparently amorphous region in which lies (by simple arithmetic) the majority of the HSV structural proteins. These include α-TIF (VP16) and the "big virion" protein (VP1).

In the earliest steps of the infectious process, the HSV glycoproteins play essential roles. Heparin sulfate proteoglycans on the surface of the cell to be infected appear able to interact with viral glycoproteins gB or gC in the initiating event. The ubiquitous nature of these proteoglycans fits with the promiscuous behavior of HSV in cell culture but not with its specificity in the animal host. Recent evidence suggests that fibroblast growth factor (FGF) may also act as an HSV receptor (KANER et al. 1990). After the binding event, gB and gD (SARMIENTO et al. 1979; JOHNSON and LIGAS 1988) become involved in fusion of the viral envelope with the plasma membrane of the cell, injecting the capsid (plus tegument) into the cell for transport to the nucleus. gH may also be part of this entry process, but its precise role is uncertain (DESAI et al. 1988). The remaining "non-essential" glycoproteins presumably also have their own special functions, since they seem unlikely to have been retained by HSV simply as immune system targets; however, these have yet to be worked out. We do know that gE and gI constitute Fc receptors and that gC is a C3b receptor (the possible relevance of these activities is discussed in Chap. 7).

Several additional HSV gene products may also turn out to be membrane proteins. These have been identified both by genetic mapping of syncytial mutants and from examination of the amino acid sequence of ORFs in the viral genome (DEBROY et al. 1985; MCGEOCH et al. 1988). Production of viral glycoproteins leads to their insertion in the nuclear membrane and in cytoplasmic and plasma membranes. Replicated viral DNA is packaged into preformed empty capsids in a process which we do not fully understand. However, encapsidation involves the "a" sequence occurring at the U_L–U_S junction and at the genomic terminals (DEISS and FRENKEL 1986) and the involvement of specific viral proteins (CHOU and ROIZMAN 1989). Other viral proteins, not included in the final form of the particle, are required for capsid assembly (PRESTON et al. 1983). Mature enveloped particles, presumably in transit from the nuclear membrane to exit from the cell, are readily seen in cytoplasmic vesicles.

The tegument proteins specifically identified thus far represent a small minority of what we believe is really there, but include some interesting activities. One of these is the virion host switchoff protein (vhs), a *γ1* protein encoded by ORF U_L41 (KWONG et al. 1988) which destabilizes and degrades host mRNA and plays a potential role in maintaining the balance between classes of viral mRNA after infection. A second tegument protein is α-TIF, present in about 1000 copies per virion and capable of activating expression from HSV α genes through the "TAATGARAT" motif (KRISTIE and ROIZMAN 1987), as we have discussed earlier. Other proteins with known functions which reside in the tegument are the α proteins. The *α4* protein was shown to be consistently present in HSV virions in small quantities (YOA and COURTNEY 1989). Recently, *α27*, but not *α0*, has also been demonstrated to be a tegument protein (COURTNEY, personal communication). The role of these regulatory proteins in the virion is undefined but they may be important in the initiation of the infectious process.

6 Gene Expression During HSV Latency

After a primary infection with HSV, the virus establishes latent infection in ganglia, from which it may be reactivated. There appear to be several structural and functional differences between virus in the latent configuration and virus during productive infection. The first of these is that the structure of the viral genome appears "endless" in latency and is likely present as extrachromosomal concatamers, probably in nucleosome-like elements (DESHMONE and FRASER 1989). There are probably 20 or fewer copies of viral DNA per latently infected neuron. The second difference between latent and productive HSV infections is that only one region of the genome (latency associated transcript; LAT) is transcriptionally active, yielding two stable (poly A−) transcripts of 2.3 and 1.8 kb. The larger unspliced species is the only form seen during productive infection (WAGNER et al. 1988). LAT transcripts are confined to the nuclei of neurons (STEVENS et al. 1987), probably because of intrinsic properties of the RNA. LAT is made from an 8.5 kb precursor RNA, and the ends of LAT are defined by splice donor and acceptor sites at the 5′ and 3′ termini, respectively.

One model for the action of LAT is that, since it is complementary to part of the *α0* gene, it might function to shut down the early stages of productive virus infection and allow latency. However, deletion mutants in LAT are able to effectively establish latency (JAVIER et al. 1988), while these same mutants fail to reactivate as efficiently as wild-type virus in the rabbit eye model (HILL et al. 1990). It could be that protein(s) from the LAT region (or from the larger transcript) help to release HSV from latency, while the LAT RNA holds back *α0* production and assists in saving the neuron from lytic infection (WAGNER 1991).

7 Concluding Remarks

We have constructed a brief picture of expression from the HSV genome, a subject which is yet to be fully elucidated. The functions of many of the more than 70 potential HSV gene products remain unexplored and several surprises presumably await us. In the context of the immune response to HSV, we have defined a number of viral proteins which serve as triggers for humoral immune responses (largely glycoproteins) and which are currently being employed in tests for the efficacy of vaccines for HSV infection. It should be pointed out that we know nothing about immune responses, humoral or cellular, against the great majority of proteins coded for by the HSV genome.

Acknowledgments. The experimental work cited here that originated from our own laboratories was supported by USPHS grants AI18449 (W.T.R. and J.H.), and AI22468 (W.T.R.) and AI25376 (J.H.). The authors wish to thank Mark Challberg, Richard Courtney, Charles Knopf, David Knipe, and Kent Wilcox for making unpublished and prepublication data available.

References

Ackermann M, Braun DK, Pereira L, Roizman B (1984) Characterization of α proteins 0, 4, and 27 with monoclonal antibodies. J Virol 52: 108–118

Banks TA, Allen EM, Dasgupta S, Sandri-Goldin R, Rouse BT (1991) Herpes simplex virus type 1 specific cytotoxic T lymphocytes recognized immediate early protein ICP27. J Virol 65: 3185–3191

Berg JM (1990) Zinc fingers and other metal-binding domains. J Biol Chem 265: 6513–6526
Booy FP, Newcomb WW, Trus BL, Brown JC, Baker TS, Steven AC (1991) Liquid-crystalline, phage-like packing of encapsidated DNA in herpes simplex virus. Cell 64: 1007–1015
Challberg MD (1986) A method for identifying the viral genes required for herpesvirus DNA replication. Proc Natl Acad Sci 83: 9094–9098
Chou J, Roizman B (1989) Characterization of DNA sequence common and DNA sequence specific proteins binding to the *cis*-acting sites for the cleavage of the terminal α sequence of herpes simplex virus 1 genome. J Virol 63: 1059–1068
Conley AF, Knipe DW, Jones PC, Roizman B (1981) Molecular genetics of herpes simplex virus. VII. Characterization of a temperature-sensitive mutant produced by in vitro mutagenesis and defective in DNA synthesis and accumulation of α polypeptides. J Virol 37: 191–206
de Bruyn Kops A, Knipe DM (1988) Formation of DNA replication structures in herpes virus-infected cells requires a viral DNA binding protein. Cell 55: 857–868
Debroy C, Pederson N, Person S (1985) Nucleotide sequence of a herpes simplex virus type 1 gene that causes cell fusion. Virology 145: 36–48
Deiss LP, Frenkel N (1986) Herpes simplex virus amplicon: cleavage of concatemeric DNA is linked to packaging and involves amplification of the terminally reiterated α sequence. J Virol 57: 933–941
Desai PJ, Schaffer PA, Minson AC (1988) Excretion of noninfectious virus particles lacking glycoprotein H by a temperature-sensitive mutant of herpes simplex virus type 1: evidence that gH is essential for virion infectivity. J Gen Virol 69: 1147–1156
Deshmone SL, Fraser NW (1989) During latency, herpes simplex virus type 1 DNA is associated with nucleosomes in a chromatin structure. J Virol 63: 943
Everett RD (1986) The products of herpes simplex virus type 1 (HSV-1) immediately early genes 1, 2, and 3 can activate HSV-1 gene expression in trans. J Gen Virol 67: 2507
Everett RD, Preston CM, Stow ND (1991) Functional and genetic analysis of the role of Vmw110 in herpes simplex virus replication. In: Wagner E (ed) Herpesvirus transcription and its regulation. CRC Press, Boca Raton. pp 50–76
Fitzgerald-Bocarsly P, Howell DM, Pettera L, Tehrani S, Lopez C (1991) Immediate early gene expression is sufficient for induction of natural killer cell indicated lysis of herpes simplex virus type 1 infection fibroblasts. J Virol 65: 3151–3160
Gaffney DF, McLaughlan J, Whitton JL, Clements JB (1985) A modular system for the assay of transcriptional regulatory signals: the sequence TAATGARAT is required for herpes simplex immediate early gene activation. Nucleic Acids Res 13: 7847–7854
Gao M, Knipe DM (1991) Potential role for HSV ICP8 DNA replication protein in stimulation of late gene expression. J Virol 65: 2666–2675
Godowski PJ, Knipe DM (1983) Mutations in the major DNA-binding protein gene of herpes simplex virus type 1 result in increased levels of viral gene expression. J Virol 47: 478–486
Gupte SS, Olson JW, Ruyechan WT (1991) The major HSV-1 DNA-binding protein is a zinc metalloprotein. J Biol Chem 266: (in press)
Hay RT, Hay J (1980) Properties of HSV-induced immediate early polypeptides. Virology 104: 230–234
Hay J, Roberts CR, Ruyechan WT, Steven AC (1987) The herpesviridae. In: Nermut M, Steven A (eds) Animal virus structure. Elsevier, New York
Hill JM, Sederati F, Javier RT, Wagner EK, Stevens JG (1990) Herpes simplex virus latent phase transcription facilitates in vivo reactivation. Virology (in press)
Holland LE, Anderson KP, Stringer JR, Wagner EK (1979) Isolation and localization of herpes simplex virus type 1 mRNA abundant before viral DNA synthesis. J Virol 31: 447–462
Holmes AM, Wietstock SM, Ruyechan WT (1988) Identification and characterization of a DNA primase activity present in herpes simplex virus type 1-infected HeLa cells. J Virol 62: 1038–1045
Homa FL, Glorioso JC, Levine M (1988) A specific 15-bp TATA box promoter element is required for expression of a herpes simplex virus type 1 late gene. Genes Dev 2: 40
Honess RW, Roizman B (1974) Regulation of herpesvirus macromolecular synthesis. I. Cascade regulation of the synthesis of three groups of viral proteins. J Virol 14: 8–19
Hope RG, Marsden HS (1983) Processing of glycoproteins induced by herpes simplex virus type 1: sulphation and nature of the oligosaccharide linkages. J Gen Virol 64: 1943–1953
Javier RT, Stevens JG, Dissette VB, Wagner EK (1988) A herpes simplex virus transcript abundant in latently infected neurons is dispensible for establishment of the latent state. Virology 166: 254
Johnson PA, Everett RD (1986) The control of herpes simplex virus type 1 late gene transcription: a TATA-box/capsite region is sufficient for fully regulated activity. Nucleic Acid Res 14: 8247

Johnson DC, Ligas MW (1988) Herpes simplex viruses lacking glycoprotein D are unable to inhibit penetration: quantitative evidence for virus-specific cell surface receptors. J Virol 62: 4605–4612
Johnson DC, Spear PG (1983) O-linked oligosaccharides are acquired by herpes simplex virus glycoproteins in the Golgi apparatus. Cell 32: 987–997
Jones KA, Yamamoto KR, Tijan R (1985) Two distinct transcription factors bind to the HSV thymidine kinase promoter in vitro. Cell 42: 559–572
Kaner RJ, Baird A, Mansukhani A, Basilico C, Summers BD, Florkiewicz RL, Hajjar DP (1990) FGF receptor is a portal of entry for herpes simplex virus type 1. Science 248: 1410–1413
Kinchington PR, Arvin AM, Ruyechan WT, Hay J (1992) The immediate early protein of VZV (IE62) is a major component of virus particles. J Virol 66: 359–366
Knipe DM (1989) The role of viral and cellular proteins in HSV regulation. Adv Virus Res 37: 85–123
Kristie TM, Roizman B (1987) Host cell proteins bind to the *cis*-acting site required for virion-mediated induction of herpes simplex virus 1 α genes. Proc Natl Acad Sci USA 84: 71–75
Kwong AD, Kruper JA, Frenkel N (1988) Herpes simplex virus virion host shutoff function. J Virol 62: 912–921
Martin S, Courtney RJ, Fowler G, Rouse BT (1988) Herpes simplex virus type 1 specific cytotoxic T lymphocytes recognize virus nonstructural proteins. J Virol 62: 2265–2273
Martin S, Zhu X, Silverstein S, Courtney RJ, Yao F, Jenkins FJ, Rouse BT (1990) Murine Cytotoxic T lymphoicytes specific for herpes simplex virus type 1 recognize the immediate early protein ICP4 but not ICPO. J Gen Virol 71: 2391–2399
Mavromara-Nazos P, Ackerman M, Roizman B (1986) Construction and properties of a viable HSV-1 recombinant lacking coding sequences for the *α47* gene. J Virol 60: 801–812
McGeoch DJ, Dalrymple MA, Davison AJ et al. (1988) The complete DNA sequence of the long unique region in the genome of herpes simplex virus type 1. J Gen Virol 69: 1531–1574
Mavromara-Nazos P, Roizman B (1987) Activation of herpes simplex virus *γ2* genes by viral DNA replication. Virology 161: 593–598
Michael N, Spector D, Mavromara-Nazos P, Kristie TM, Roizman B (1988) The DNA binding properties of the major regulatory protein *α4* of herpes simplex viruses. Science 239: 1531–1534
Olivo PD, Challberg MD (1991) Functional analysis of the HSV gene products involved in DNA replication. In: Wagner E (ed) Herpesvirus transcription and its regulation. CRC Press, Boca Raton, pp 137–150
Poffenberger KL, Roizman B (1985) Studies on non-inverting genome of a viable herpes simplex virus 1. Presence of head-to-tail linkages in packaged genomes and requirements for circularization after infection. J Virol 53: 589–595
Post LE, Mackem S, Roizman B (1981) The regulation of α genes of herpes simplex virus: expression of chimeric genes produced by fusion of thymidine kinase with α gene promoters. Cell 24: 555–565
Preston VG, Coates AM, Rixon FJ (1983) Identification and characterization of a herpes simplex virus gene product required for encapsidation of viral DNA. J Virol 45: 1056–1064
Roizman B (1990a) Herpesviridae: a brief introduction. In: Fields B, Knipe DM (eds) Virology. Raven, New York, pp 1787–1794
Roizman B (1990b) Whether herpesviruses? In: Lopez C, Mori R, Roizman B, Whitley RT (eds) Immunobiology and prophylaxis of human herpesvirus infections. Plenum, New York, pp 285–292
Roizman B, Roane PR Jr (1964) Multiplication of herpes simplex virus. II. The relation between protein synthesis and the duplication of viral DNA in infected HEp-2 cells. Virology 22: 262–269
Roizman B, Sears A (1990) Herpes simplex viruses and their replication. In: Fields B, Knipe DM (eds) Virology. Raven, New York, pp 1795–1841
Sandri-Goldin R (1991) Analysis of the regulatory activities of the HSV-1 protein ICP27. In: Wagner E (ed) Herpesvirus transcription and its regulation. CRC Press, Boca Raton, pp 77–104
Sarmiento M, Haffey M, Spear PG (1979) Membrane proteins specified by herpes simplex viruses. III. Role of glycoprotein VP7(B2) in virion infectivity. J Virol 29: 1149–1158
Schrag JD, Prasad BVV, Rixon FJ, Chiu W (1989) Three-dimensional structure of the HSV1 nucleocapsid. Cell 56: 651–660
Sears AE, Halliburton IW, Meignier B, Silver S, Roizman B (1985) Herpes simplex virus mutant deleted in the *α22* gene: growth and gene expression in permissive and restrictive cells, and establishment of latency in mice. J Virol 55: 338–346
Sekulovich RE, Leary K, Sandri-Goldin RM (1988) The herpes simplex virus type 1 α protein ICP27 can act as a trans-repressor or a trans-activator in combination with ICP4 and ICPO. J Virol 62: 4510

Silver S, Roizman B (1985) Thymidine kinase chimeras are identically transcribed but regulated as $\gamma 2$ genes in herpes simplex virus genomes and as β genes in cell genomes. Mol Cell Biol 5: 518–528

Stevens JG, Wagner EK, Devi-Rao GB, Cook ML, Feldman LT (1987) RNA complementary to a herpesvirus α gene mRNA is predominant in latently infected neurons. Science 235: 1056

Stringer J, Holland JL, Swanstrom R, Pivo K, Wagner E (1979) Quantitation of herps simplex virus type 1 RNA in infected HeLa cells. J Virol 21: 889–901

Su L, Knipe DM (1987) Herpes simplex α protein ICP27 can inhibit or augment viral gene *trans*-activation. Virology 176: 496–504

Wagner EK (1991) Herpesvirus transcription: general aspects. In: Wagner E (ed) Herpesvirus transcription and its regulation. CRC Press, Boca Raton, pp 1–16

Wagner EK, Flanagan WM, Devi-Rao G, Zhang YF, Hill JM, Anderson KP, Stevens JG (1988) The herpes simplex virus latency associated transcript is spliced during the latent phase of infection. J Virol 62: 4577

Weller S (1991) Genetic analysis of HSV genes required for genome replication. In: Wagner E (ed) Herpesvirus transcription and its regulation. CRC Press, Boca Raton, pp 105–136

Yao F, Courtney R (1989) A major transcriptional regulatory protein of HSV-1 is associated with purified virions. J Virol 63: 3338–3344

Pathogenesis of Herpes Simplex Virus Infection and Animal Models for Its Study

L. R. STANBERRY

1 Introduction

Overcoming infection with herpes simplex virus (HSV) depends upon a complex interaction between virus and host. This interaction most commonly produces asymptomatic infection, as illustrated by the high prevalence of HSV antibodies in individuals with no history of clinically recognized illness (GIBSON et al. 1990; JOHNSON et al. 1989). These findings suggest that host responses are generally effective in limiting viral disease. Clearly this is not always the case since millions of individuals experience symptomatic HSV infections with disease ranging from trivial to life threatening (WHITLEY 1990). From epidemiological studies and clinical observations we have learned that both viral and host factors may influence whether an individual experiences an asymptomatic infection or is afflicted with severe herpetic disease. Clinical studies, however, are inherently limited and do not permit detailed exploration of the host and viral determinants of disease expression. The use of well-characterized experimental animal models has proven invaluable in further defining the natural history of HSV

Division of Infectious Diseases, Children's Hospital Research Foundation, University of Cincinnati College of Medicine, Cincinnati, Ohio, USA

Current Topics in Microbiology and Immunology, Vol. 179

infection. With the recent advances in molecular biology and immunology, animal models are being increasingly used to investigate the role of selected viral genes and specific immune responses in the pathogenesis and immunobiology of HSV infection. This chapter will review the experimental animal models used to study HSV and summarize our current understanding of the pathogenesis of infection, with a primary focus on genital herpes. The immunobiology of HSV, as it pertains to viral pathogenesis and animal model research, will receive minimal attention in this chapter but will be presented in greater detail in other chapters in this volume.

2 HSV Infections in Humans—An Overview

HSV type 1 (HSV-1) and type 2 (HSV-2) are closely related viruses that produce primary and recurrent infections. Two properties of these viruses are their ability to replicate in skin and mucous membranes producing vesicular lesions and to replicate in neural tissues causing neurologic symptoms. The pattern of clinical disease resulting from primary HSV infection is largely determined by the portal of entry of the virus and the immune status of the host. Disease is generally limited to the proximity of the portal of entry and to neural tissues innervating the inoculation site. Dissemination of the infection is rare except in the immunologically immature or incompetent host. Recovery from the primary infection is associated with the establishment of viral latency. Reactivation of latent virus can result in recurrent herpetic infections, which are usually less severe and generally limited to the same anatomic location as the primary infection. Two important exceptions to this generalization are recurrent

Table 1. Herpes simplex virus infections

Portal of entry	Immune status	Initial infections	Common reactivation infection	Rare reactivation infection
Oropharynx	Competent	Stomatitis or pharyngitis	Herpes labialis	Herpes encephalitis
Eye	Competent	Kerato-conjunctivitis	Herpes keratitis	Herpes encephalitis
Genitalia	Competent	Primary genital herpes	Recurrent genital herpes	
Abraded skin	Competent	Herpes gladi-atorum	Cutaneous herpes	
Cuticle	Competent	Herpetic whitlow	Herpetic whitlow	
Skin/eye/mouth(?)	Immature	Neonatal herpes	Cutaneous herpes	Herpes encephalitis
Any of the above	Compromised	Disseminated herpes	Localized severe herpes	Disseminated infection

infections in the immunosuppressed patient, which can be severe and occasionally result in disseminated disease, and herpes encephalitis in the normal host, which is produced by the reactivation of latent virus and results in devastating neurologic injury. The types of infections produced by HSV are shown in Table 1. The symptoms produced by HSV-1 and HSV-2 are so similar that clinical findings in an individual patient cannot be used to reliably distinguish between these two viruses. Nevertheless, clinical studies have noted biological and epidemiological differences in the two virus types. For example, infants surviving neonatal HSV-2 infection are more likely to be neurologically impaired than those infected with HSV-1 (WHITLEY et al. 1991). This observation suggests that HSV-2 may be more neurovirulent than HSV-1. Another difference noted is in the ability of the two virus types to produce recurrent infections. A patient with HSV-2 genital infection is twice as likely to have recurrent disease as an individual with HSV-1 genital infection, and the recurrences are eight to ten times more frequent (COREY et al. 1983; REEVES et al. 1981). In contrast, HSV-1 is more likely to produce recurrent orolabial infections than HSV-2 (COREY and SPEAR 1986). These observations suggest that the incidence and frequency of recurrent infections are influenced by virus type. The biological basis for the differences in the behavior of the two virus types is currently under investigation.

3 Requirements for a Good Animal Model

The definition of a good animal model is largely influenced by the questions to be addressed by the proposed research. For studies designed to explore the natural history of an infection, the most important requirement is that the disease produced in the animal closely mimics the disease observed in humans. Ideally this includes not only the clinical manifestations of the infection but also its pathogenesis and immunobiology. However, some animal models have very little resemblance to human disease but are useful to explore specific aspects of viral pathogenesis. Thus, selection of an animal model will be determined by the type of research to be conducted. For example, the intraneuronal transport of virus has been studied using the mouse footpad inoculation model (COOK and STEVENS 1973). This model uses an unnatural route of inoculation but, because the sciatic nerve, which innervates the footpad, can be removed easily for examination, this model has proven popular. Inoculation of mice by a variety of routes results in fatal illness while HSV infections in humans are generally self-limited. Despite this major difference, lethal challenge models have proven very useful in screening putative antiviral drugs (FIELD 1988; HSIUNG and CHAN 1989; KERN et al. 1978). Recurrent herpetic infections are common in humans but are not usually observed in experimental models, a notable exception being the guinea pig model of genital herpes in which both spontaneous and induced recurrent disease may occur.

Table 2. Factors to consider in selecting an experimental animal model

1. Does the infection in the animal mimic the infection in humans?
   What is the outcome of primary infection?
   Is a latent infection established?
   Does the animal experience recurrent infections?
2. What is the route of inoculation?
   Natural route (e.g., intravaginal)?
   Unnatural route (e.g., intraperitoneal)?
3. Are the host responses to infection similar to those of humans?
   Availability of characterized immunologic reagents?
   Availability of defined inbred strains?
4. Can the animal be used to predict antiviral drug efficacy?
   What is the clinical endpoint? Death? Reduction in disease severity?
   What are the pharmacokinetic comparisons?
5. Is the model useful in predicting vaccine efficacy?
   What is the measure of efficacy? Protection from death or from disease?
   Does animal show immunologic responses to immunization similar to those of humans?
6. How convenient is the model to use?
   Cost to purchase the animals?
   Cost to house the animals?
   General availability of the species?
   Ease of handling the species?
   Reproducibility of the model?

Consequently, this model has been employed to investigate the natural history of recurrent infections (SCRIBA and TATZBER 1981; STANBERRY 1986, 1989). The choice of an animal model may also be influenced by the availability of defined inbred animal strains and well-characterized immunologic reagents. Thus, most studies of the immunobiology of HSV infection have utilized mouse models (LOPEZ 1975; ROUSE et al. 1988). Animal models that utilize a natural route of inoculation were developed in order to study specific types of HSV disease. For example, corneal inoculation of mice, guinea pigs, and rabbits is used to investigate herpes keratitis (NESBURN et al. 1972; STULTING et al. 1985; WANDER et al. 1987), while intravaginal inoculation of a variety of animals (mice, hamsters, guinea pigs, marmosets, aotus monkeys) is useful to study genital herpes (MEIGNIER et al. 1990; OVERALL et al. 1975; RENIS 1975; STANBERRY et al. 1982). Factors that should be considered in selecting an animal model are summarized in Table 2.

4 Animal Models Useful in the Study of HSV Pathogenesis

4.1 Models of HSV Ocular Disease

Historically, the rabbit has been a favored experimental animal in studies of herpetic eye disease (GOODPASTURE and TEAGUE 1923; NESBURN et al. 1972). Infection by corneal scarification has been shown to produce corneal and stromal disease, neovascularization, and ulcerative blepharitis. Intraneuronal HSV transport through sensory nerves has been demonstrated, and virus has been shown to establish latent infection in the trigeminal ganglia. One major strength of the model is that rabbits experience both spontaneous and induced recurrent infections (GREEN et al. 1981; STROOP and SCHAEFER 1987). The guinea pig has also been used to study ocular herpes (SMOLIN et al. 1976; WANDER et al. 1987). Corneal infection in the guinea pig is similar to that observed in the rabbit and human, however, recurrent infections have not been described. This model has been useful to explore the pathogenesis of latent infection (TENSOR and HSIUNG 1977). One potential advantage of the guinea pig compared to the rabbit model is the availability of inbred animal strains. Over the past two decades mouse models of ocular disease have become increasingly popular. Both corneal and intraocular inoculation models have been developed (STULTING et al. 1985; WHITTUM et al. 1984). Besides the obvious advantages of size and cost, the mouse models have the added advantages of defined inbred animal strains and well-characterized immunologic reagents. Consequently, these models are currently the most popular to study the immunology and immunopathology of HSV ocular disease, both keratitis and retinitis (see the Chapter by DUBIN et al., this volume; WHITTUM et al. 1984), and to investigate immunotherapeutic approaches to the control of herpetic ocular disease (RECTOR et al. 1982). The mouse ocular model is also finding favor with molecular biologists interested in exploring the molecular basis of viral pathogenesis (POLVINO-BODNAR et al. 1987; SPIVACK and FRASER 1987).

4.2 Models of Central Nervous System Infection

Mouse models of central nervous system (CNS) infection are used primarily to study intraneuronal virus transport and neurovirulence. Transport studies generally use footpad inoculation and follow movement of virus from the periphery through nerves to sensory ganglia, spinal cord, and finally cerebral cortex (COOK and STEVENS 1973; KLEIN and DESTEFANO 1983; KRISTENSSON et al. 1971). Such studies have established a major tenet of HSV pathogenesis: virus spreads from the site of inoculation via intraneuronal axoplasmic transport rather than simply cell to cell or via a hematogenous route. Studies designed to explore HSV neurovirulence have utilized mice inoculated by either the footpad or intracranial routes (DIX et al. 1983). These experiments often compare the lethality of parental and mutant viruses in order to explore the role of selected viral genes (or regions of the viral genome) in HSV neurovirulence (CHOU et al. 1990; JAVIER et al. 1988; THOMPSON et al. 1989). Experiments using the mouse footpad model demonstrated that infection of an animal with two avirulent viruses could result in recombination of the viruses in vivo to produce a virulent (and lethal) recombinant viral strain (JAVIER et al. 1986). This important observation not only provides insight into how new virus strains might arise, but also illustrates a theoretical disadvantage to prototype live, genetically engineered vaccine viruses that might recombine in vivo to produce virulent recombinants. One of the most important recent discoveries regarding HSV pathogenesis was made using the mouse footpad model. It has long been debated whether persistent HSV infection represented a low-level productive infection that was below the threshold of detection by conventional techniques or, whether the virus persisted in a nonreplicating latent state. Studies exploring the transcription of the viral genome in latently infected mouse sensory ganglion cells showed that only a limited region of the HSV genome was transcribed during latent infection (STEVENS et al. 1987). Hence, HSV persists in a nonreplicating state with only minimal transcriptional activity during latent infection. The role of the latency associated transcripts in viral pathogenesis is an area of intensive investigation. While these and related murine models have been instrumental in exploring viral determinants important in neuronal infection, the models have been less useful in studying the pathogenesis of herpes encephalitis. Inoculation of mice by any number of routes results in an acute infection which involves the entire cerebral cortex. Herpes encephalitis in humans (other than neonates) is generally restricted to the temporal lobe and appears to result from reactivation of latent virus in 50% of affected individuals (NAHMIAS et al. 1982). An experimental infection more like that observed in humans has recently been described in rabbits (STROOP and SCHAFFER 1986). Animals were inoculated by either direct HSV injection of the olfactory bulb or by application of the virus onto the nasal mucosa. Virus was shown to move along the olfactory tract to the frontal and temporal regions of the rabbit brain producing a localized but generally fatal infection of the CNS. As with humans, some animals were subclinically infected. Immuno-suppression of the subclinically infected rabbits resulted in reactivation of latent virus and the production of an encephalitis which was restricted to the temporal lobes. This model should prove useful in further exploring the pathogenesis and control of herpes encephalitis.

4.3 Models of Neonatal Infection

HSV infection of the human newborn is a complex, life threatening disease (WHITLEY et al. 1991). Infection may be limited to the portal of entry (skin, eye, or mouth), localized to the CNS, or disseminate to involve multiple organ systems. The factors important in either limiting the infection or contributing to its dissemination are unknown. Mouse models of neonatal infection have been developed using young (1–21 day old) animals inoculated with HSV-1 or HSV-2 by a variety of routes. Animals inoculated intranasally develop both CNS and pulmonary infections. This route of infection has been used in the evaluation of experimental antiviral chemotherapy (KERN et al. 1986). The potential value of postexposure serum prophylaxis was shown in mice infected subcutaneously (BARON et al. 1976). Guinea pigs were also used to study neonatal infection. Corneal inoculation of newborn (2–4 day old) guinea pigs produced encephalitis in 30% of animals and 83% of the survivors were shown to be latently infected (TENSER and HSIUNG 1971). More recent studies have shown that 97% of newborn (less than 24 h) outbred Hartley guinea pigs develop skin, eye, or mucous membrane disease between 2 and 7 days following intranasal inoculation with HSV-2. Progression of the disease to include CNS and/or multiple organ involvement was noted and approximately half of the animals died within 4–11 days postinoculation. As occurs with many infected human newborns, surviving guinea pigs exhibited spontaneous herpetic recurrences of the skin, eye, or mucous membranes (mouth or nares). Acyclovir was shown to reduce mortality but, as with humans, did not prevent recurrences in the survivors (BRAVO et al. 1990). The guinea pig has also been used to explore the effect of age on neonatal infection (MANI et al. 1991). This model appears to share many features with neonatal infection in humans including a natural route of inoculation, the development of both localized and disseminated disease, and spontaneous recurrent infections in the survivors. Since the guinea pig has a long pregnancy (67 days) and a placenta similar to that of the human, this model may prove to be useful to explore maternal/perinatal factors that influence outcome. Similar guinea pig models can be used to study the effect of pregnancy on the immunobiology of related herpesvirus (e.g., cytomegalovirus) infections (HARRISON and MYERS 1990).

4.4 Models of Genital Infection

As HSV became an increasingly common sexually transmitted pathogen, several animal models were developed to facilitate investigation of the pathophysiology and control of genital herpes. Like humans, these models used a natural route of inoculation (intravaginal) and the animals exhibited clinical evidence of vaginitis. However, unlike humans, most species, including mice, hamsters, marmosets, and owl monkeys, died, generally of an ascending encephalomyelitis (KATZIN et al. 1967; OVERALL et al. 1975; RENIS 1975). A notable but expensive exception was the Cebus monkey, which developed a nonfatal infection that was clinically, virologically, and serologically similar to that of humans (NAHMIAS et al. 1971; REEVES et al. 1976). A more practical species, the guinea pig, also developed a self-limited infection following intravaginal HSV inoculation (reviewed in STANBERRY 1986). This model shares many of the features of genital herpes in humans. As illustrated in Fig. 1, replication of the virus occurs in the genital tract resulting in high titers of HSV in vaginal secretions. After a 3–4 day "incubation" period, animals begin to develop herpetic lesions on the external genital skin (perineum) (Fig. 2). Initially, these lesions are discrete vesicles but, by 5–8 days after inoculation, the lesions progress to confluent vesiculoulcerative lesions. Hemorrhagic crusts (eschars) cover the ulcers by 8–10 days with loss of crusts and complete healing by 13–16 days post-HSV inoculation. The clinical course of genital skin disease in the guinea pig is shown in Fig. 2. As with humans, guinea pigs also establish a latent infection

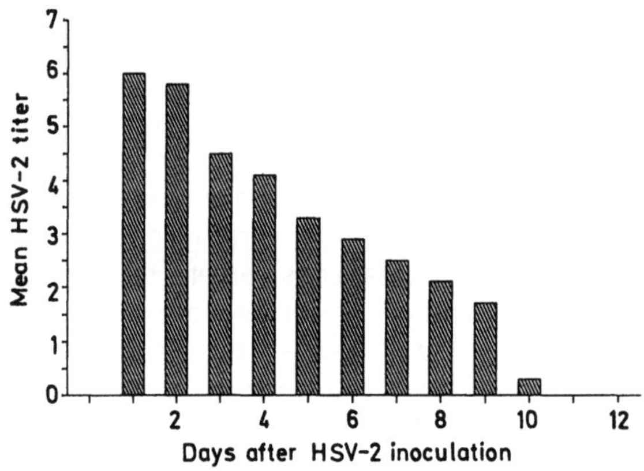

Fig. 1. The course of HSV-2 replication in female guinea pigs. Mean titer (pfu/ml) of virus in vaginal swab samples

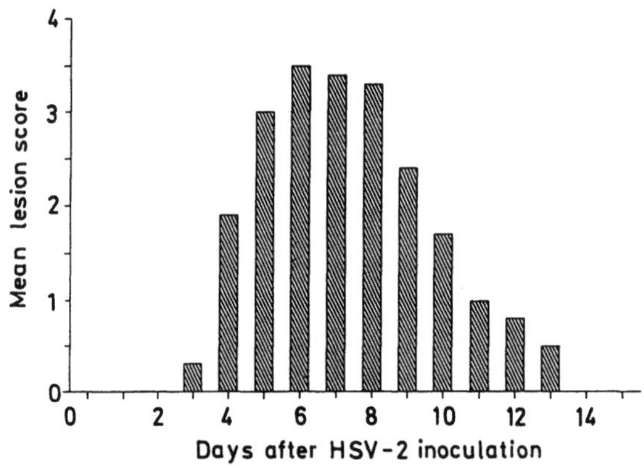

Fig. 2. The clinical course of genital HSV-2 skin disease in female guinea pigs. Severity of disease was assessed by the following score: *0*, no disease; *1+*, redness or swelling; *2+*, a few small vesicles; *3+*, several large vesicles; *4+*, several large ulcers. A declining score was used during the healing stage

in sacral dorsal root ganglia (SCRIBA and TATZBER 1981; STANBERRY et al. 1982) and develop recurrent HSV infections. Like humans, these recurrences may be manifested as either asymptomatic viral shedding from the genital tract or clinically apparent lesions on the genital skin (STANBERRY 1986). Most recurrent infections in humans occur spontaneously and the same is true for the guinea pig. Recurrent perigenital infections in humans can be induced by ultraviolet radiation (PERNA et al. 1987; WHEELER 1975). Likewise, ultraviolet radiation has been used to induce recurrent genital HSV infections in guinea pigs (STANBERRY 1989). Interestingly, both HSV-1 and HSV-2 produce indistinguishable primary infections in humans and guinea pigs, but HSV-2 is more likely to cause recurrent infections in both species (COREY and SPEAR 1986; STANBERRY et al. 1985b). Numerous immunologic responses to genital HSV infection in the guinea pig have also been described. These include neutralizing and ELISA antibody responses, mucosal antibody responses, plasma interferon (IFN) activity and in vitro interleukin-2 (IL-2) production, antibody-dependent cell-mediated cytotoxicity, peripheral blood mononuclear cell (PBMC) proliferation and PBMC-mediated cytolytic responses, natural killer cell activity, and lymphokine activated killer cell responses. This model has been used to explore viral pathogenesis (HSIUNG et al. 1984; STANBERRY 1986), evaluate antiviral chemotherapy (BERNSTEIN et al. 1986; FRASER-SMITH et al. 1983), study the effects of viral mutations on virulence and latency (ANDERSON et al. 1980; STANBERRY et al. 1985a), investigate the immunobiology of genital infection (HARRISON et al. 1988), examine the neural spread of virus in vivo (STANBERRY 1990a), and develop prototype vaccines for possible clinical use (MEIGNIER et al. 1988; STANBERRY et al. 1987). One important observation arising from experiments using the guinea pig model of genital herpes was that immunization of latently infected animals with HSV subunit vaccines could reduce the frequency and severity of recurrent infections (HO et al. 1989; STANBERRY et al. 1988). For more than 50 years clinical researchers had hypothesized that immunotherapy of patients experiencing frequent herpetic recurrences should result in improved control of recurrent infections. Despite numerous clinical trials with live or inactivated HSV vaccines, no convincing evidence could be generated to support the concept (reviewed in STANBERRY 1990b, 1991). One major confounder in such studies was the significant placebo effect. With the development of a model of spontaneous recurrent infections, in which placebo effects could be controlled, it became possible to rigorously evaluate the concept of immunotherapy of persistent viral infections. The studies using the guinea pig have shown that treatment of latently infected animals with HSV glycoprotein vaccines does reduce recurrent infections. This effect is significantly influenced by the choice of adjuvants and routes of immunization (STANBERRY et al. 1989). Initial studies have shown that treatment with glycoproteins also results in an enhanced IL-2 response and a persistent increase in HSV-specific cytolytic responses. Further studies of the mechanism of glycoprotein immunotherapy are under way (BERNSTEIN et al. 1991). As a closing comment on the guinea pig, it is interesting to note that a male model of genital herpes has also been reported (STEPHANOPOULOS et al. 1989).

4.5 Miscellaneous Models of HSV Infection

It is an impossible task to list all the other animal models of HSV infection that have been developed. However, at least a few deserve brief mention. Cutaneous infection of the mouse ear has been an important model used to explore the pathogenesis of induced recurrent infections (HILL et al. 1975; BLYTH et al. 1976) and the zosteriform spread of virus (BLYTH et al. 1984). Cutaneous infection of the guinea pig was also used to study zosteriform HSV spread (BERNSTEIN and

STANBERRY 1986) and to evaluate topical antiviral drug therapy (SPRUANCE and MCKEOUGH 1988). A guinea pig footpad model has been used to study the immunobiology of cutaneous recurrent infections (DONNENBERG et al. 1980). A recently developed rat model was employed to investigate acute and latent ganglionic infection (BLONDEAU et al. 1991). Experimental inoculation of the tree shrew, a primitive primate, results principally in an infection of the liver and spleen, this unusual infection has been used to explore the organotropism of latent HSV (DARAI et al. 1978; ROSEN-WOLFF et al. 1989). For those concerned about the ethics of animal research but still in need of an in vivo system, there is the human model of HSV infection. Remarkable reports have included experimentally inoculating the gingival mucosa of a 12 month old infant in order to establish HSV as the etiological agent of gingivostomatitis or the intravenous administration of HSV to 11 cancer patients in order to study HSV viremia (BLACK 1942; GELLER et al. 1953).

5 The Pathogenesis of HSV Infection

The pathogenesis of HSV infection, as illustrated by genital herpes, is depicted in Fig. 3. In the susceptible host primary infection begins with the transmission of HSV to the genital mucosa. At least three HSV glycoproteins, gB, gD, and gH, are involved in virus binding and penetration of cells (CAI et al. 1988; DESAI 1988; FULLER and SPEAR 1987). The cellular attachment site appears to be heparan sulfate, a sulfated glycosaminoglycan (WUDUNN and SPEAR 1989), although other membrane components, such as the fibroblast growth factor receptor may also play a role in penetration of virus into the cell (KANER et al.

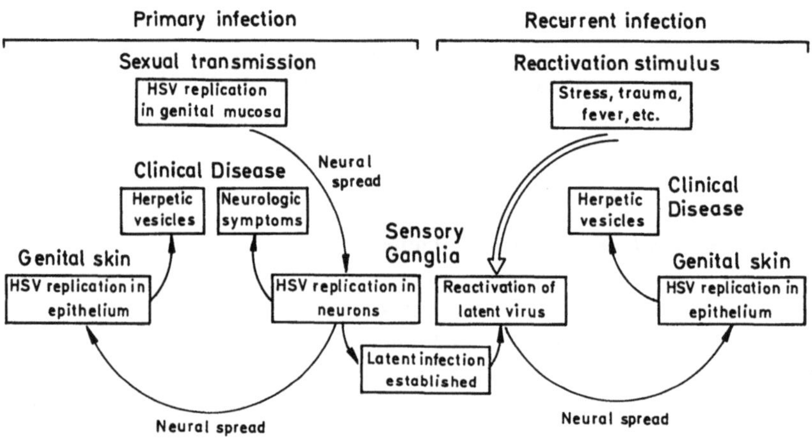

Fig. 3. The pathogenesis of primary, latent, and recurrent genital herpes. (From STANBERRY 1991)

1990). Penetration can be prevented by cytochalasins, suggesting that a change in the cell membrane's cytoskeletal structure occurs with virus attachment that triggers a microfilament activity which mediates internalization of the virus (ROSENTHAL et al. 1988). The virus loses its envelope upon entry into the cytoplasma and the naked nucleocapsid migrates into the nucleus where replication of the HSV DNA occurs (LYCKE et al. 1984; ROIZMAN and SEARS 1990). Progeny virus are assembled in the nucleus and acquire an envelope by budding through the inner nuclear membrane (SCHWARTZ and ROIZMAN 1969). Posttranslational modification of the viral glycoproteins occurs as the enveloped virus moves through the Golgi apparatus. The mature virus is then transported by a monensin-sensitive process from the Golgi apparatus to the cell surface where release of the newly replicated virus takes place (JOHNSON and SPEAR 1982). In women the cervix is presumed to be the main site of the primary infection (YEN et al. 1965). Local replication of virus occurs in epithelial cells, continuing for 1–3 weeks and producing peak titers of approximately 10^5 per vaginal swab specimen in the first 1–3 days of the infection (BROWN et al. 1979). Virus also rapidly enters sensory nerves and moves by axoplasmic transport to the nuclei of sacral dorsal root ganglia (COOK and STEVENS 1973; KRISTENSSON et al. 1971; STANBERRY et al. 1982). Replication of the virus in the sensory ganglia results in a neuritis which may be responsible for the severe pain experienced by many individuals with genital herpes (COREY et al. 1983). HSV may also ascend to the spinal cord where further replication may result in myelitis or aseptic meningitis (KLASTERSKY et al. 1972; VON HOFF et al. 1975). Following replication in sensory neurons HSV is transported through capsaicin-sensitive unmyelinated nerves back to the genital skin where further replication results in the characteristic lesions of primary genital herpes (STANBERRY 1990a). The intraneuronal movement of virus through this neural arc appears necessary for the development of clinically apparent HSV disease (STANBERRY 1986).

To control the primary infection the host produces a variety of humoral, cytokine, and cellular responses. These responses act quickly to limit viral replication resulting in a linear decline in the recovery of virus from the genital tract (BROWN et al. 1979; STANBERRY et al. 1982). Suppression of immune responses results in a more severe infection (MULLER et al. 1972; SIMMONS and NASH 1987), while augmentation of host immunity facilitates control of disease (HARRISON et al. 1988; STANBERRY et al. 1987). (Detailed discussion of the various host immune responses engendered by HSV infection is provided in other chapters in this volume.) Control of the primary infection by the immune system, however, does not result in eradication of all HSV. Some virus continues to persist in a nonreplicating latent state within the nuclei of sensory ganglia (CROEN et al. 1991). Virus may also persist in a latent state in vaginal and cervical tissues, although the specific cell type harboring the virus is unknown and the biological significance of extraganglionic viral latency is uncertain (CLEMENTS and SUBAK-SHARPE 1988; STANBERRY et al. 1985a; WALZ et al. 1977). The events involved in the establishment and maintenance of the latent state are also unclear. A limited region of the viral genome is transcriptionally active.

The function(s) of these latency associated RNAs are unknown but do not appear essential for either the establishment or maintenance of the latent state (SEDARTI et al. 1989; STEVENS et al. 1987). Latent virus may be reactivated to an infectious (replicating) form which in turn may produce recurrent infections. A variety of stimuli, such as ultraviolet radiation, have been demonstrated to induce recurrent infections (PERNA et al. 1987; STANBERRY 1989) by triggering the reactivation of latent virus (N. BOURNE and L. STANBERRY, unpublished data). The mechanism responsible for reactivation is unknown but is suspected to involve changes in factors important in the regulation of neuronal gene transcription. Once reactivated, the infectious HSV is transported via capsaicin-sensitive sensory nerve fibers back to the skin where further replication in epithelial cells produces the characteristic vesicular lesions of recurrent HSV infections. The frequency of recurrent infections declines with time (MYERS et al. 1975; STANBERRY et al. 1985b). Likewise, the concentration of latent HSV DNA present in dorsal root ganglia also declines over time (B. CONNELLY and L. STANBERRY, unpublished data). This observation suggests that recurrent infections do not "recharge" the pool of latent virus in sensory ganglia. Whether the reduced number of recurrences seen with aging is a consequence of less latent virus being available for reactivation or is due to a failure of aging neurons to be able to reactivate the latent virus is unknown. An interesting aspect of recurrent HSV infections is how they occur despite a full range of anti-HSV host immune responses. This failure of immune surveillance is not understood and is an area of ongoing scientific endeavor.

6 Conclusion

HSV-1 and HSV-2 are large DNA viruses that share approximately 50% nucleic acid homology and code for at least 70 proteins. The function of most of these proteins is unknown; however, many are not required for replication in cell culture. Hence, much of the HSV genome appears to code for proteins that provide the virus with specialized functions such as neurovirulence, latency, and reactivation. Studies using well-characterized animal models will permit further definition of the genes and gene products essential for the pathogenesis of HSV infection in vivo. Likewise, animal model studies will be required to develop a clinically meaningful understanding of host responses important in the control of primary and recurrent HSV infections. The ultimate goal of such research efforts will be to develop new strategies for the prevention and treatment of HSV-induced disease.

Acknowledgments. This work was supported in part by grants RO1-AI22667, ROI-AI29687 and NO1-AI15101 from the National Institutes of Health. I would like to thank my colleagues, Drs. Beverly Connelly, Nigel Bourne, and Fernando Bravo for their help with the illustrations and for thoughtful discussion.

References

Anderson CA, August MJ, Hsiung GD (1980) Pathogenecity of wild-type and temperature sensitive mutants of herpes simplex virus type 2 in guinea pigs. Infect Immun 30: 159–169

Baron S, Worthington MG, Williams J, Gaines JW (1976) Postexposure serum prophylaxis of neonatal herpes simplex virus infection of mice. Nature 261: 505–506

Bernstein DI, Stanberry LR (1986) Zosteriform spread of herpes simplex virus type 2 genital infection in the guinea-pig. J Gen Virol 67: 1851–1857

Bernstein DI, Stanberry LR, Harrison CJ, Kappes J, Myers MG (1986) Effects of oral acyclovir treatment of initial genital HSV-2 infection upon antibody response, recurrence patterns and subsequent HSV-2 reinfection in guinea pigs. J Gen Virol 67: 1601–12

Bernstein DI, Harrison CJ, Jenski LJ, Myers MG, Stanberry LR (1991) Cell mediated immunological responses and recurrent genital herpes in the guinea pig: effects of glycoprotein therapy. J Immunol 146: 3571–3577

Black WC (1942) The etiology of acute infectious gingivostomatitis. J Pedriatr 20: 145–160

Blondeau JM, Aoki FY, Glavin GB, Nagy JI (1991) Characterization of acute and latent herpes simplex virus infection of the dorsal root ganglia in rats. Lab Anim 25: 97–105

Blyth WA, Hill TJ, Field HJ, Harbour DA (1976) Reactivation of herpes simplex virus infection by ultraviolet light and possible involvement of prostglandins. J Gen Virol 33: 547–550

Blyth WA, Harbour DA, Hill TJ (1984) Pathogenesis of zosteriform spread of herpes simplex virus in the mouse. J Gen Virol 65: 1477–1486

Bravo FJ, Myers MG, Bernstein DI, Connelly BL, Harrison CJ, Kier AB, Stanberry LR (1990) Neonatal HSV infection in guinea pigs: a model to test antivirals. Abstracts of the 30th Intersci Conf Antimicrob Agents Chemother, Atlanta, abstract no 1104

Brown ZA, Kern ER, Spruance SL, Overall JC Jr (1979) Clinical and virological course of herpes simplex genitalis. West J Med 130: 414–421

Cai WH, Gu B, Person S (1988) Role of glycoprotein B of herpes simplex virus type 1 in viral entry and cell fusion. J Virol 62: 2596–2604

Chou J, Kern ER, Whitley RJ, Roizman B (1990) Mapping of herpes simplex virus-1 neurovirulence to gamma 34.5, a gene nonessential for growth in culture. Science 250: 1262–1266

Clements GB, Subak-Sharpe (1988) Herpes simplex virus type 2 establishes latency in the mouse footpad. J Gen Virol 69: 375–383

Cook ML, Stevens JG (1973) Pathogenesis of herpetic neuritis and ganglionitis in mice: evidence for intra-axonal transport of infection. Infect Immun 7: 272–288

Corey L, Spear PG (1986) Infections with herpes simplex viruses. N Engl J Med 314: 749–757

Corey L, Adams HG, Brown ZA, Holmes KK (1983) Genital herpes simplex virus infection: clinical manifestations, course and complications. Ann Intern Med 98: 958–972

Croen KD, Ostrove JM, Dragovic L, Straus SE (1991) Characterization of herpes simlex virus type 2 latency associated transcription in human sacral ganglia and in cell culture. J Infect Dis 163: 23–28

Darai G, Schwaier A, Komitowski D, Munk K (1978) Experimental infection of Tupaia belanger (tree shrews) with herpes simplex virus types 1 and 2. J Infect Dis 137: 221–226

Desai PJ, Schaffer PA, Minson AC (1988) Excretion of non-infectious virus particles lacking glycoprotein H by a temperature sensitive mutant of herpes simplex virus type 1: evidence that gH is essential for virion infectivity. J Gen Virol 69: 1147–1156

Dix RD, McKendall RR, Baringer JR (1983) Comparative neurovirulence of herpes simplex virus type 1 strains after peripheral or intracerebral inoculation of BALB/c mice. Infect infect 40: 103–112

Donnenberg AD, Chaikof E, Aurelian L (1980) Immunity of herpes simplex virus type 2: cell-mediated immunity in latently infected guinea pigs. Infect Immun 30: 99–109

Field HJ (1988) Animal models in the evaluation of antiviral chemotherapy. In: Field HJ (ed) Antiviral agents: the development and assessment of antiviral chemotherapy, vol I. CRC Press, Boca Raton, pp 67–84

Fraser-Smith EB, Smee DF, Matthews TR (1983) Efficacy of the acyclic nucleoside 9-(1,3-dihydoxy-2-propoxymethyl) guanine against primary and recrudescent genital herpes simplex virus type 2 infections in guinea pigs. Antimicrob Agents Chemother 24: 883–887

Fuller AO, Spear PG (1987) Anti-glycoprotein D antibodies that permit adsorption but block infection by herpes simplex virus type 1 prevent virion-cell fusion at the cell surface. Proc. Natl Acad Sci USA 84: 5454–5458

Geller P, Coleman VR, Jawetz E (1953) Studies on herpes simplex virus: V. The fate of viable herpes simplex virus administered intravenously to man. J Immunol 71: 410–418
Gibson JJ, Hornung CA, Alexander GR, Lee FK, Potts WA, Nahmias AJ (1990) A cross-sectional study of herpes simplex virus types 1 and 2 in college students: occurrence and determinants of infection. J Infect Dis 162: 306–312
Goodpasture EW, Teague O (1923) Transmission of the virus of herpes febrilis along nerves in experimentally infected rabbits. J Med Res 44: 139–184
Green MT, Rosborough JP, Dunkle EC (1981) In vivo reactivation of herpes simplex virus in rabbit trigeminal ganglia: electrode model. Infect Immun 34: 69–76
Harrison CJ, Myers MG (1990) Relation of maternal CMV viremia and antibody response to the rate of congenital infection and intrauterine growth retardation. J Med Virol 31: 222–228
Harrison CJ, Jenski L, Voychehovski T, Bernstein DI (1988) Modification of immunological responses and clinical disease during topical R-837 treatment of genital HSV-2 infection. Antiviral Res 10: 209–224
Hill TJ, Field HJ, Blyth WA (1975) Acute and recurrent infection with herpes simplex virus in the mouse—a model for studying latency and recurrent disease. J Gen Virol 28: 341–353
Ho RJY, Burke RL, Merigan TC (1989) Antigen-presenting liposomes are effective in treatment of recurrent herpes simplex virus genitalis in guinea pigs. J Virol 63: 2951–2958
Hsiung GD, Chan VF (1989) Evaluation of new antiviral agents: II. The use of animal models. Antiviral Res 12: 239–258
Hsiung GD, Mayo DR, Lucia HL, Landry ML (1984) Genital herpes: pathogenesis and chemotherapy in the guinea pig model. Rev Infect Dis 6: 33–50
Javier RT, Sedafati F, Stevens JG (1986) Two avirulent herpes viruses interact in the animal to generate a virulent recombinant and a lethal infection. Science 234: 746–748
Javier RT, Izumi KM, Stevens JG (1988) Location of herpes simplex virus neurovirulence gene dissociated from high-titer virus replication in the brain. J Virol 62: 1381–1387
Johnson DC, Spear PG (1982) Monensin inhibits the processing of herpes simplex virus glycoproteins, their transport to the cell surface, and the egress of virions from infected cells. J Virol 43: 1102–1112
Johnson RE, Nahmias AJ, Magder LS, Lee FK, Brooks CA, Snowden CB (1989) A seroepidemiological survey of the prevalence of herpes simplex virus type 2 infection in the United States. N Engl J Med 321: 7–12
Kaner RJ, Baird A, Mansukhani A, Basilico C, Summers BD, Florkiewciz RZ, Hajjar DP (1990) Fibroblast growth factor receptor is the portal of cellular entry for herpes simplex virus type 1. Science 248: 1410–1413
Katzin DS, Connor JD, Wilson LA, Sexton RS (1967) Experimental herpes simplex infection in the owl monkey. Proc Soc Exp Biol Med 125: 391–398
Kern ER, Richards JT, Overall JC Jr, Glasgow LA (1978) Alteration of mortality and pathogenesis of three experimental herpesvirus hominis infections of mice with adenine arabinoside 5′-monophosphate, adenine arabinoside, and phosphonoacetic acid. Antimicrob Agents Chemother 13: 53–60
Kern ER, Richards JT, Overall JC Jr (1986) Acyclovir treatment of disseminated herpes simplex virus type 2 infection of weanling mice: alteration of mortality and pathogenesis. Antiviral Res 6: 189–195
Klastersky J, Cappel R, Snocck JM, Flament J, Thiry L (1972) Ascending myelitis in association with herpes-simplex virus. N Engl J Med 287: 182–184
Klein J, DeStefano E (1983) Dissemination of herpes simplex virus in ganglia after footpad inoculation in neurectomized and non-neurectomized mice. Acta Virol 77: 231–238
Kristensson K, Lycke E, Sjostrand J (1971) Spread of herpes simplex in peripheral nerves. Acta Neuropathol (Berl) 17: 44–53
Lopez C (1975) Genetics of natural resistance to herpesvirus infections in mice. Nature 258: 152–153
Lycke E, Kristensson K, Svennerholm B, Vahine A, Ziegler R (1984) Uptake and transport of herpes simplex virus in neurites of rat dorsal root ganglia cells in culture. J Gen Virol 65: 55–64
Mani C, Stanberry LR, Bravo FJ, Bernstein DI, Myers MG (1991) Neonatal herpes simplex virus infection in guinea pigs: effect of age on clinical disease and outcome (abstract) Pediatr Res 29: 255A
Meignier B, Longnecker R, Roizman B (1988) In vivo behavior of genetically engineered herpes simplex viruses R7017 and R7020: construction and evaluation in rodents. J Infect Dis 158: 602–614

Meignier B, Martin B, Whitley J, Roizman B (1990) In vivo behavior of genetically engineered herpes simplex viruses R7017 and R7020. II. Studies in immunocompetent and immunosuppressed owl monkeys (*Aotus trivirgatus*). J Infect Dis 162: 313–321

Muller SA, Hermann FC, Winkelman RK (1972) Herpes simplex infections in hematological malignancies. Am J Med 52: 102–114

Myers MG, Oxman MN, Clark JE, Arndt KA (1975) Failure of neutral-red photodynamic inactivation in recurrent herpes simplex infections. N Engl J Med 293: 945–949

Nahmias AJ, London WT, Catalona LW, Fuccillo DA, Sever JL, Graham C (1971) Genital herpesvirus hominis type 2 infection: an experimental model in Cebus monkeys. Science 171: 297–298

Nahmias AJ, Whitley RJ, Vistine AN, Takei Y, Alford CA Jr, the NIAID Collaborative Antiviral Study Group (1982) Herpes simplex encephalitis: laboratory evaluations and their diagnostic significance. J Infect Dis 145: 829–836

Nesburn AB, Cook ML, Stevens JG (1972) Latent herpes simplex virus: isolation from rabbit trigemmalganglia between episodes of recurrent ocular infection. Arch Ophthalmol 88: 412–417

Overall JC Jr, Kern ER, Schlitzer RL, Friedman SB, Glasgow LA (1975) Genital herpesvirus hominis infection in mice. I. Development of an experimental model. Infect Immun 11: 476–480

Perna JJ, Mannix ML, Rooney JF, Notkins AL, Straus SE (1987) Reactivation of latent herpes simplex virus infection by ultraviolet light: a human model. J Am Acad Dermatol 17: 473–478

Polvino-Bodnar M, Orberg PK, Schaffer PA (1987) Herpes simplex virus type 1 ori_L is not required for virus replication or for the establishment and reactivation of latent infection in mice. J Virol 61: 3528–3535

Rector JT, Lausch RN, Oakes JE (1982) Use of monoclonal antibodies for analysis of antibody-dependent immunity to ocular herpes simplex virus type 1 infection. Infect Immun 38: 168–174

Reeves WC, DiGiacomo RF, Alexander ER, Lee CK (1976) Latent herpesvirus hominis from the trigeminal and sacral dorsal root ganglia of Cebus monkeys. Proc Soc Exp Biol Med 153: 258–261

Reeves WC, Corey L, Adams HG, Vontver LA, Holmes KK (1981) Risk of recurrence after episodes of genital herpes: relation of HSV type and antibody response. N Engl J Med 305: 315–319

Renis HE (1975) Genital infection of female hamsters with herpesvirus hominis type 2 (HVH-2). Proc Soc Exp Biol Med 150: 723–727

Roizman B, Sears AE (1990) Herpes simplex viruses and their replication. In: Field BN, Knipe DM (eds) Virology. Raven, New York, pp 1795–1841

Rosenthal KS, Roess D, Barisas BG (1988) Herpes simplex virus type 1 initiates mobilization of cell surface proteins. Biochim Biophys Acta 942: 38–44

Rosen-Wolff A, Scholz J, Darai G (1989) Organotropism of latent herpes simplex virus type 1 is correlated to the presence of a 1.5 kb RNA transcript mapped within the BamHI DNA fragment B (0.738 to 0.809 map units). Virus Res 12: 43–52

Rouse BT, Norley S, Martin S (1988) Antiviral cytotoxic T lymphocyte induction and vaccination. Rev Infect Dis 10: 16–33

Schwartz J, Roizman B (1969) Similarities and differences in the development of laboratory strains of herpes simplex virus in HEp-2 cells: electron microscopy. J Virol 4: 879–889

Scriba M, Tatzber F (1981) Pathogenesis of herpes simplex virus infections in guinea pigs. Infect Immun 34: 655–661

Sedarti F, Izumi KM, Wagner EK, Stevens JG (1989) Herpes simplex virus type 1 latency-associated transcription plays no role in the establishment or maintenance of a latent infection in murine sensory neurons. J Virol 63: 4455–4458

Simmons A, Nash AA (1987) Effect of B cell suppression on primary infection and reinfection of mice with herpes simplex virus. J Infect Dis 155: 469–475

Smolin G, Okumoto MA, Ohno S (1976) Experimental herpetic keratitis in the guinea pig. Surv Ophthalmol 21: 205–208

Spivack JG, Fraser NW (1987) Detection of herpes simplex virus type 1 transcripts during latent infection in mice. J Virol 61: 3841–3847

Spruance SL, McKeough MB (1988) Evaluation of antiviral treatments for recurrent herpes simplex labialis in the dorsal cutaneous guinea pig model. Antiviral Res 9: 295–313

Stanberry LR (1986) Herpesvirus latency and recurrence. Prog Med Virol 33: 61–77

Stanberry LR (1989) Animal model of ultraviolet-radiation-induced recurrent herpes simplex virus infection. J Med Virol 28: 125–128

Stanberry LR (1990a) Capsaicin interfers with the centrifugal spread of virus in primary and recurrent genital herpes simplex virus infection. J Infect Dis 162: 29–34

Stanberry LR (1990b) Subunit viral vaccines: therapeutic and prophylactic uses. In: Aurelian L (ed) Herpesviruses, the immune systems and aids. Kluwer, Boston, pp 309–341
Stanberry LR (1991) Herpes simplex virus vaccines. Semin Pediatr Infect Dis 2(3): 178–185
Stanberry LR, Kern ER, Richards JT, Abbott TM, Overall JC Jr (1982) Genital herpes in guinea pigs: pathogenesis of the primary infection and description of recurrent disease. J Infect Dis 146: 397–404
Stanberry LR, Kit S, Myers MG (1985a) Thymidine kinase-deficient herpes simplex virus type 2 genital infection in guinea pigs. J Virol 55: 322–328
Stanberry LR, Kern ER, Richards JT, Overall JC Jr (1985b) Recurrent genital herpes simplex virus infection in the guinea pig. Intervirology 24: 226–231
Stanberry LR, Bernstein DI, Burke RL, Pachl C, Myers MG (1987) Recombinant herpes simplex virus glycoproteins: protection against initial and recurrent genital herpes. J Infect Dis 155: 914–920
Stanberry LR, Burke RL, Myers MG (1988) Herpes simplex virus glycoprotein treatment of recurrent genital herpes. J Infect Dis 157: 156–163
Stanberry LR, Harrison CJ, Bernstein DI, Burke RL, Shukla R, Ott G, Myers MG (1989) Herpes simplex virus glycoprotein immunotherapy of recurrent genital herpes: factors influencing efficacy. Antiviral Res 11: 203–214
Stephanopoulos DE, Myers MG, Bernstein DI (1989) Genital infections due to herpes simplex virus type 2 in male guinea pigs. J Infect Dis 159: 89–95
Stevens JG, Wagner EK, Devi-Rao GB, Cook ML, Feldman LT (1987) RNA complementary to a herpesvirus alpha gene mRNA is prominent in latently infected neurons. Science 235: 1056–1059
Stroop WG, Schaefer DC (1986) Severity of experimentally reactivated herpetic eye disease is related to the neurovirulence of the latent virus. Invest Opthalmol Vis Sci 28: 229–237
Stulting RD, Kindle JC, Nahmias A (1985) Patterns of herpetic keratitis in inbred mice. Invest Ophthalmol Vis Sci 26: 1360–1367
Tenser RB, Hsiung GD (1977) Pathogenesis of latent herpes simplex virus infection of the trigeminal ganglion in guinea pigs: effects of age, passive immunization, and hydrocortisone. Infect Immun 16: 69–74
Thompson RL, Rogers SK, Zerhusen MA (1989) Herpes simplex virus neurovirulence and productive infection of neural cells is associated with a function which maps between 0.82 and 0.832 map units on the HSV genome. Virology 172: 435–450
Von Hoff DD, Luckey M, Wallace J (1975) Herpes simplex virus type 2 meninigitis following herpes progenitalis. West J Med 123: 490–491
Walz MA, Price RW, Hayashi K, Katz BJ, Notkins AL (1977) Effect of immunization on acute and latent infections of vaginouterine tissue with herpes simplex viruses types 1 and 2. J Infect Dis 135: 744–752
Wander AH, Bubel HC, McDowell SG (1987) The pathogenesis of herpetic ocular disease in the guinea pig. Arch. Arch Virol 95: 197–209
Wheeler CE Jr (1975) Pathogenesis of recurrent herpes simplex infections. J Invest Dermatol 65: 341–346
Whitley RJ (1990) Herpes simplex viruses. In: Fields BN, Knipe DM (eds) Virology. Raven, New York, pp 1843–1887
Whitley R, Arvin A, Prober C, Corey L, Burchett S, Plotkin S, Starr S, Jacobs R, Powell D, Nahmias A, Sumaya C, Edwards K, Alford C, Caddell G, Soong S-J (1991) Predictors of morbidity and mortality in neonates with herpes simplex virus infections. N Engl J Med 324: 450–454
Whittum JA, McCulley JP, Niederkorn JY, Streilein JW (1984) Ocular disease induced in mice by anterior chamber inoculation of herpes simplex virus. Invest Ophthalmol Vis Sci 25: 1065–1971
WuDunn D, Spear PG (1989) Initial interaction of herpes simplex virus with cells is binding to heparan sulfate. J Virol 63: 52–58
Yen SSC, Reagan JW, Rosenthal MS (1965) Herpes simplex infection in the female genital tract. Obstet Gynecol 25: 479–492

The Role of Immune Mechanisms in Control of Herpes Simplex Virus Infection of the Peripheral Nervous System

A. SIMMONS, D. TSCHARKE, and P. SPECK

Institute of Medical and Veterinary Science, Frome Road, Adelaide, South Australia 5000

Current Topics in Microbiology and Immunology, Vol. 179

1 Introduction

Underlying the recurrent cutaneous lesions caused by herpes simplex virus (HSV) is a sophisticated virus-host relationship involving primary sensory neurons. These cells may become productively infected, but sometimes virus replication is interrupted for long periods, producing a reservoir of latent herpes in the nervous system from which infection can periodically reactivate. How the balance is tipped in favor of either productive or latent infection and how productive infection of neurons is rapidly controlled (which, in general, it is) are prominent contemporary questions for herpes virologists. In this chapter, we critically review the clinical and experimental evidence suggesting that the host's adaptive immune response makes a vital contribution to the control of established HSV replication in the nervous system, and we dwell on the controversial issue of the fate of productively infected neurons.

2 The Neurodermatome

The peripheral nervous system (PNS) of vertebrates is segmentally organized and bilaterally symmetrical. Two dorsal sensory nerve roots and two ventral motor roots arise from each segment of the cerebrospinal axis. On each side, a dorsal and ventral root combine to form a sensorimotor nerve trunk, which leaves the spinal column and becomes the conduit through which many skeletal muscles are controlled and via which sensory information from the periphery is relayed to the central nervous system (CNS). The cell bodies of primary sensory neurons are located in swellings (ganglia) on the dorsal roots (Fig. 1). Each cell is unipolar, i.e., originates a single fiber, which divides at a T-junction close to the cell body, sending one branch into the CNS via the nerve root and another to the periphery along the spinal nerve. Physiological and morphological studies define several distinct populations of primary sensory neurons, but our understanding of the relationship between structure and function is at a rudimentary stage. Collection of sensory neurons into dorsal root ganglia (DRG) on a segmental basis would appear to be for housekeeping purposes because, unlike sympathetic ganglia, the DRG are not sites of intraneural communication; individual neuronal soma are isolated by a surround of capsular cells which are thought to be peripheral counterparts of CNS microglial cells. Gathering of the sensory neurons into the ganglia is imprecise and stray neurons are commonplace in the dorsal root and proximal peripheral nerve.

The patch of skin innervated by a dorsal root ganglion is termed a dermatome, and, although the distribution of sensory fibers in the skin remains largely segmental, successive dermatomes of the trunk overlap their neighbors by at least 50%. A corollary is that a small patch of skin contains sensory nerve fibers arising from at least three adjacent ganglia. For detailed and elegant descriptions of the dermatomes of humans, the free penicillate nerve endings

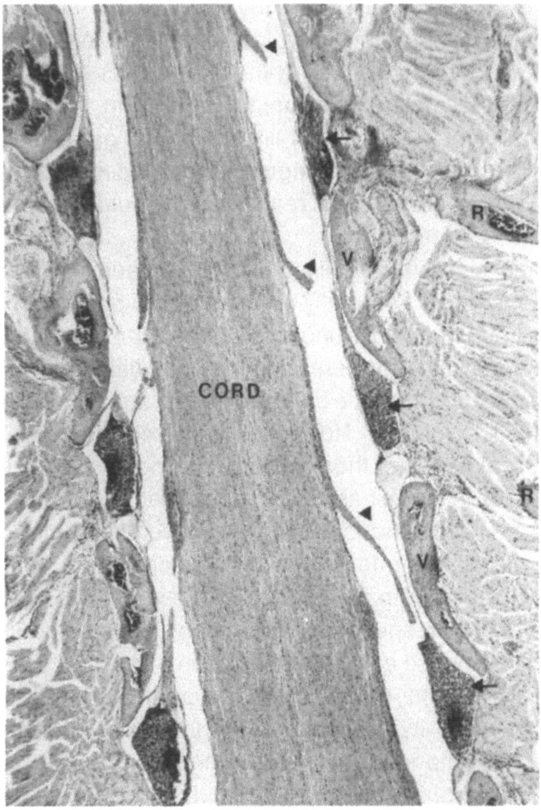

Fig. 1. Coronal section of the lower thoracic region of the spinal column of a mouse, illustrating spinal cord, dorsal nerve roots (*arrowheads*), dorsal root ganglia (*arrows*), vertebrae (*V*), and ribs (*R*)

of the human skin, and the innervation of the cornea, the reader is referred to FOERSTER (1933), CAUNA (1973), and WHITEAR (1960), respectively. The dermatome, the sensory ganglion, and the spinal nerve and its dorsal root comprise the *neurodermatome*. This functional unit is sufficient to propagate HSV and provides an environment for lifelong persistence of the virus within the host.

3 Experimental Systems

3.1 Animal Models for Investigating the Control of PNS Infection

3.1.1 Desired Characteristics of Model Systems

It has been our aim to develop and refine a model specifically for studying the control of HSV infection within the nervous system. The ideal model has four basic characteristics: First, following inoculation of HSV into skin or mucous membrane there must be a reproducible, self-limited, primary infection during which virus is translocated intra-axonally to sensory ganglia (KRISTENSSON et al. 1971; COOK and STEVENS 1973), resembling the spread of virus in humans (BARINGER 1975). Second there should be a self-limited acute ganglionic infection. Third, latency should be stable and indistinguishable at the molecular level from latency in humans. Fourth, spontaneous reactivation should give rise to recrudescent lesions.

Guinea pigs probably come closest to fulfilling all four criteria. Herpes simplex recrudesces spontaneously and frequently in this species, particularly after inoculation of the vagina (reviewed by HSIUNG et al. 1984). Consequently, guinea pigs have been of enormous value for assessing the potential influence of immunomodulators, including vaccines, on the course of reactivated infection (see the chapter by STANBERRY, this volume). However, the inbred laboratory mouse is a powerful immunological tool, and there is merit in using this species for studying control of earlier stages of the disease. Reproducible acute ganglionic infection and stable latency are readily induced in mice by several peripheral routes of inoculation, and the courses of infection in lumbosacral and upper cervical ganglia after inoculation of the footpad (WILDY 1967; STEVENS and COOK 1971) and ear pinna (HILL et al. 1975), respectively, have been documented in detail. For more than a decade the mouse ear model has been the benchmark by which other models for examining anti-herpes immunity, particularly those aspects relating to delayed hypersensitivity, are judged. Only the fourth characteristic of the ideal model, namely, spontaneous reactivation and recrudescence, is a stumbling block in mice, which in our judgement does not reduce the value of the species for studying innate and adaptive responses to primary infection. We therefore selected inbred mouse models for further development.

3.1.2 Zosteriform Spread of HSV

In 1923 TEAGUE and GOODPASTURE observed that rabbits injected in the midflank with HSV developed a band-like lesion several days later that resembled herpes zoster. They concluded that the infection had spread throughout the entire neurodermatome, the final stage being centrifugal flow of virus down the spinal nerve trunk to produce the zosteriform rash. Zosteriform lesions have subsequently been induced in guinea pigs (STANBERRY et al. 1982), rats (TANAKA and SOUTHAM 1965), and mice (SIMMONS and NASH 1984) and are occasionally seen in humans, particularly in association with primary infection (MOK 1971; MUSIC et al. 1971). Although perhaps not widely appreciated, this indicates that primary HSV infection at any peripheral site might have a local zosteriform component superimposed on or adjacent to initial local replication; in support of this hypothesis, experimental denervation of the skin reduces virus yield from developing primary lesions in mice (SIMMONS, unpublished).

Zosteriform spread is a recognized feature of infection at several sites, including the ear pinna which shares innervation with skin of the neck (BLYTH

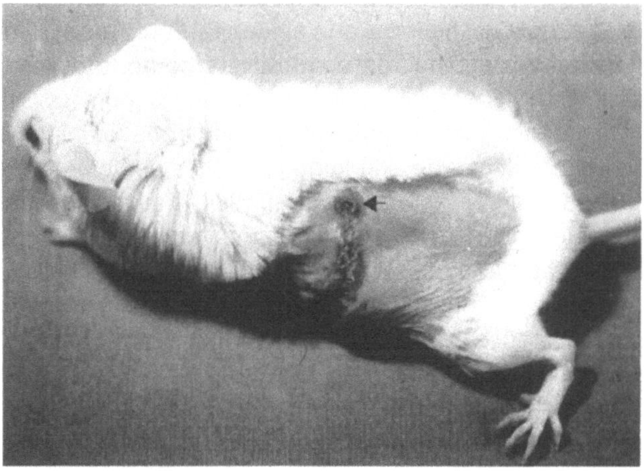

Fig. 2. Zosteriform lesion in the tenth thoracic dermatome of a BALB/c mouse 5 days after inoculating the left flank with 2×10^5 pfu of HSV-1, strain *SC16* (FIELD et al. 1979) at the point shown by the *arrow*

et al. 1984). However, by developing our model in the midflank region (SIMMONS and NASH 1984), we have been able to make maximal and novel use of the well preserved segmental innervation of the thorax and abdomen (Fig. 2), allowing spread of virus to the PNS, and then from segment to segment within the PNS, to be charted with extreme precision (SIMMONS and NASH 1987; SPECK and SIMMONS 1991), particularly when coronal sections are used (Fig. 1). Key aspects of the current system, which is highly reproducible with respect to virus recovery and lesion development, are as follows: (1) A precisely defined area of the flank is infected by scarification of the skin; subcutaneous or intradermal injection of virus is much less efficient at inducing zosteriform lesions unless chemical irritants are applied, as in early reports (TEAGUE and GOODPASTURE 1923). (2) Productive infection is centralized on the tenth thoracic somatic (vertebral) segment. (3) By careful mouse strain selection, different degrees of innate attenuation of the infection can be achieved (SIMMONS and LA VISTA 1989). (4) The flank region provides for simple and rapid surgical access to the nerve trunks of each segment, allowing the cutaneous and neuronal components of the neurodermatome to be disconnected from one another at any stage of the infection. (5) As zosteriform lesions erupt, both the emergence of virus from nerve endings and subsequent infection of normal epidermal cells provide a powerful model with which to study the formation of recrudescent lesions (SIMMONS and NASH 1984, 1985).

3.2 In Vitro Systems

There is merit in studying the effects on neuronal infection of each component of the immune system in isolation, and the most elegant prospect of achieving this goal is provided by recent attempts to recreate the neurodermatome in vitro using cultured human fetal ganglionic neurons and epidermal cells. The system is based on the ability of neurites extending from cultured fetal neurons to pierce a diffusion barrier and enter a separate chamber within the culture vessel, allowing HSV (or other substances such as antibodies, interferons, or immune cells) to be applied exclusively to either the extending tip of the neurite or to the neuronal bodies (ZIEGLER and HERMAN 1980; LYCKE et al. 1984). The most recent refinement is the addition of fetal epidermal cells to the chamber into which the neurites grow. The extending neural processes interact with epidermal cells in a manner that morphologically resembles that seen in human fetal skin at 30 weeks of gestation (CUNNINGHAM 1991). At the time of writing, the intraaxonal passage of virus between the culture chambers of this system is being studied intensively, with the future prospect of applying immunoregulatory substances to one or the other chamber (CUNNINGHAM, personal communication).

4 Interaction of HSV with the PNS

4.1 Acute Infection

Inoculation of a sublethal dose of virus into the skin of experimental animals characteristically leads, 2–4 days later, to an acute ganglionic infection that typically resolves after a further 3–5 days (reviewed by WILDY et al. 1982). Virus travels to the PNS via the nerve cylinder (WILDY 1967), apparently parasitizing the fast axonal flow process (LYCKE et al. 1984). Indeed, HSV is a potentially powerful tracer of neurological pathways (BAK et al. 1977, 1978; UGOLINI et al. 1987). Productive infection within the sensory ganglia is limited almost entirely to neurons; capsular cells may express viral antigens but electron microscopic studies suggest that infection within such cells is abortive (DILLARD et al. 1972; COOK and STEVENS 1973). Schwann cells in the nerve trunk may, however, produce characteristic herpesvirus particles (RABIN et al. 1968; SEVERIN and WHITE 1968). While these observations in animal models have been important in formulating our current concept of herpesvirus pathogenesis, there is no proof that human sensory neurons become productively infected during uncomplicated primary disease; however, there is also no evidence to the contrary.

4.2 Latent Infection

Although primary and recurrent infection within the nervous system may infrequently give rise to rapidly fatal encephalitis, the outcome that usually predominates in humans is latency, in which the viral genome is sequestered in a nonreplicating state (reviewed by STEVENS 1989). Mutant viruses that are unable to initiate the cascade of gene expression associated with productive infection, because of lesions in immediate early genes (LEIB et al. 1989; KATZ et al. 1990) or absence of functional α-transinducing factor in the virion (STEINER et al. 1990), retain the ability to establish latency, suggesting that the molecular pathways of productive and latent infection are divergent from the outset. Superficially, this implies that an abortive productive infection is not responsible for persistence of the genome, but the strikingly large amount of viral DNA present in latently infected tissue (CABRERA et al. 1980; ROCK and FRASER 1983; EFSTATHIOU et al. 1986) casts doubt on this conclusion because replication of HSV DNA is probably dependent on prior synthesis of several virally encoded polypeptides (reviewed by KNIPE 1989). The controversial issue of the fate of infected neurons and whether the adaptive immune response has a role in persistence of the viral genome is examined in detail in Sect. 7.

4.3 Reactivation

In experimental systems HSV can be reactivated from ganglia in vitro or in vivo, resulting in the production of infectious virus particles by neurons (PLUMMER et al. 1970; STEVENS and COOK 1971; HARBOUR et al. 1981, 1983). In humans and some animal models reactivation occurs spontaneously despite the presence of well developed immunity. Although it does not prevent reactivation or even, in some instances, recrudescence, it is likely that the immune system plays an important role in rapidly controlling these events because: (a) reactivation only rarely leads to discernible neurological damage and (b) immunosuppression enhances the severity of recrudescences. The fate of neurons in which HSV reactivates has been the subject of intense controversy for decades. Reactivation can be surgically induced (CARTON and KILBOURNE 1952; WALZ et al. 1974), which in mice causes an apparent reduction in the number of latently infected sites, implying that reactivated cells are killed by the ensuing productive infection (MCLENNAN and DARBY 1980). In contrast, a holistic view of the process of recrudescence seems to suggest that production of infectious virus progeny by neurons cannot rapidly induce cytolysis, as it does in cultured cells, because it is necessary for virions to be transported to the periphery after their assembly, which requires the complex and energy-dependent fast axonal transport process to be intact. Further, some individuals suffer many hundreds of recrudescences within a small area of a single dermatome during their lifetime without demonstrable loss of sensation. Primary sensory neurons are a finite resource and we find it most unlikely that there is sufficient redundancy within the sensory nervous system to allow for death of even a single neuron during each reactivation.

5 The Influence of Host Genes on the Severity of Neurological Infection

5.1 MHC-Linked Genes

Many parameters determine the severity of primary neural infection with HSV, including virus strain (FIELD and WILDY 1978; RICHARDS et al. 1981), non-adaptive host responses, and the efficacy of adaptive immunity. Presumably many, if not all, of these also influence the outcome of reactivation events. Against a background of so many interacting factors there is little prospect of identifying host genetic determinants of disease severity in naturally infected human populations unless specific genes are targeted. In this context the polymorphic, yet well characterized, major histocompatibility complex (MHC: H-2 in mouse, HLA in humans) is an obvious target because MHC phenotype is readily identifiable and the region is of interest owing to its immunoregulatory functions. MHC glycoproteins play a pivotal role in restricting the cellular immune response because they are required for presentation of antigens, in the form of peptides, to T lymphocytes (MARRACK and KAPPLER 1987) and there is a rigorous association between T cell phenotype and the restricting MHC class, such that $CD8^+$ and $CD4^+$ cells recognize antigen in the context of class I and class II MHC gene products, respectively (SWAIN 1983). Several studies have attempted to correlate frequency of recrudescence with MHC (HLA) type in humans and some (predictably) weak associations have been reported (RUSSELL and SCHLAUT 1977; GALLINA et al. 1989). It is unfortunate that there is no satisfactory murine model of recurrent disease because this precludes the application of inbred mouse strains, particularly MHC congenic strains, to the study of recrudescence.

The situation is reversed for primary infection, which is not readily amenable to immunogenetic analysis in humans but for which numerous murine models are available. Consequently, genetic resistance to experimental primary herpes is well established particularly with respect to non-MHC-linked genes that convey powerful nonadaptive resistance to mice with the C57BL background (LOPEZ 1975). Until recently there were no reports suggesting that MHC genes might influence the quality of the protective response to this infection. However, the experimental protocols used in the majority of studies have been weighted against the discovery of such genetic influences because resistance has generally been equated with survival of the animal after intraperitoneal injection of virulent virus, which rapidly leads, in susceptible mice, to an overwhelming infection of the nervous system before specific immunity has developed. Using congenic mice with the resistant C57BL background, it has been possible to demonstrate a substantial influence of H-2 genes on the extent of primary neurological infection (SIMMONS 1989). Following inoculation of HSV into the skin of mice, the amount of virus recovered from sensory ganglia varies significantly between H-2 congenic strains. Differences become apparent 7 days after infection, at which time the severity of disease in C57B10.Br mice (H-2$K^k/^kD^k$) is two to three orders of magnitude greater than that in C57B10.D2 (H-2$K^dI^dD^d$) animals. The natural intra-H-2 *k/d* recombinant strain, B10A (H-2$K^kI^kD^d$), behaves in an intermediate fashion suggesting that class I loci alone can influence recovery, implicating $CD8^+$ T cells in the response. The in vitro proliferative response of human T-cells to HSV glycoprotein B and its homologues in other human herpesviruses is coinfluenced by HLA type, further supporting the hypothesis that MHC genes affect the pathobiology of the disease (CHAN et al. 1989) and also suggesting that infection with other herpesviruses might alter the pathogenesis of herpes simplex. The observed substantial variation in the ability of inbred H-2 congenic mice to control HSV infection of the PNS underlines the importance of genetic diversity with respect to the MHC in outbred populations, and challenge with infectious agents could, of course, be an important selective force in maintaining such diversity. More important in the present context is the inescapable conclusion that the efficacy of adaptive immune responses may be an important determinant of the severity of ganglionic infection.

5.2 Non-MHC-Linked Genes

Animals of the C57BL background survive intraperitoneal challenge with more than 1000 times the dose endured by a variety of other inbred mouse strains (LOPEZ 1975). This potent resistance is controlled by two autosomal dominant non-H-2 genetic loci (LOPEZ 1980), but the location of these genes and the identity of the proteins which they encode remain unknown. It is unlikely that resistance is caused by lack of a functional virus receptor because this is normally incompatible with dominant inheritance. While the intraperitoneal model has yielded valuable information, it does not mimic the progression of virus from the skin to the PNS which occurs during natural infection. Consequently this and other systemic models of infection are not readily applicable to the study of factors that influence the extent of PNS disease. In contrast, potentially valuable information has been gained from mouse hybrid analysis of resistance to acute ganglionic infection following various peripheral routes of inoculation (PRICE and SCHMITZ 1979; SIMMONS and LA VISTA 1989). In the flank zosteriform model (SIMMONS and NASH 1984), C57BL10 mice delay the onset of detectable neurological infection by up to 2 days compared with other strains. Like resistance to lethal challenge, this trait is dominantly inherited with no evidence of H-2 or sex linkage, but detailed Mendelian type analysis indicates complex genetic control of neural infection (SIMMONS and LA VISTA 1989), unlike the relatively straightforward host genetics of the intraperitoneal model. Segregation of resistance among F1 backcrossed and F2 hybrid mice generated from C57BL10 (resistant) and A/J (susceptible) parents suggests that four loci are involved, although the exact number of genes cannot be deduced with confidence because too many assumptions about the relative dominance of the various alleles would be required.

Novel approaches will be needed for further analysis of the multifactorial nature of resistance to neural infection; predictably, a useful strain distribution pattern cannot be found in the CXB recombinant inbred (RI) set of mice and the herpesvirus resistance gene *Cmv-1* does not appear to convey resistance to HSV-1 (Scalzo et al. 1990). Some CXB RI strains may differ at only one of the genetic loci of interest; F1 backcross and F2 analyses using selected CXB RI parents is a novel and untested strategy that might identify such a circumstance, if reproducible phenotypic differences can be found.

The mechanism by which non-MHC-linked genes convey resistance is unknown. Restriction of virus at the level of the PNS is manifested at an early stage of the infection, 3–5 days after inoculating the flank, suggesting that nonadaptive mechanisms are likely to be responsible, concordant with the lack of MHC involvement. However, IgG-bearing cells have been detected in the spinal ganglia of C57BL6 mice as early as 4 days after infection of the footpad (Cook and Stevens 1983), so it is not inconceivable that a rapid adaptive response could contribute to what otherwise appears to be innate resistance. Also worthy of note is the ability of C57BL mice to produce at least tenfold more interferon (IFN)-α/β than BALB/c animals within 4 h of infection (Zawatzky et al. 1982a), and genetic control of this early IFN response bears some resemblance to that of restricted PNS infection (Zawatzky et al. 1982b). Exchanging the IFN-α structural gene clusters of C57BL6 and BALB/c mice does not produce a corresponding reversal of IFN responses (DeMaeyer-Guignard et al. 1986), implying that regulatory loci rather than structural genes are responsible for the trait. One such locus, *If-1*, was transferred from the BALB/c to the C57BL background during the construction of C57BL6.H-28 congenic mice; however, this strain displays a fully resistant phenotype in the flank zosteriform model, suggesting that *If-1* has no significant influence on neurological infection with HSV (Simmons, unpublished).

6 Antibody

6.1 Control of Primary Infection

6.1.1 Passive Transfer Experiments: What Do They Tell Us?

Many investigators have demonstrated that polyclonal antiserum or monoclonal antibodies administered early in the course of primary infection can limit the spread of virus to the sensory ganglia (reviewed by Simmons and Nash 1985). High titers are required to achieve this effect, and the efficacy of antibody administration begins to decline 24–48 h after infection, which coincides with the time at which virus arrives in the sensory ganglia (Cook and Stevens 1973; McKendall et al. 1979). The mechanism by which early administration of immune serum reduces the severity of ganglionic infection is not known, but it is probable that a high titer of antibody in the tissues before or during the first few hours after infection results in partial neutralization of the inoculum and decreased uptake of virus by nerves. SETHI (1983) reported enhanced development of H-2-restricted cytotoxic T lymphocytes (CTLs) by passive transfer of antibodies 12 h after infection, raising the possibility that administration of antibody modulates other responses. Under natural conditions antibody is not detectable in the infected host until several days after inoculation of the skin, by which time ganglionic infection is established. Furthermore, administration of antibody so as to produce physiological titers fails to protect animals, even when given early (ENNIS 1973; OAKES 1975), unless disease has been enhanced by sublethal immunosuppression (WORTHINGTON et al. 1980). Therefore the recognized protective effect of early transfer of antiserum does not help us understand the natural role of antibody molecules in controlling infection in the nervous system but does serve to demonstrate the potential power of antiserum as a prophylactic agent against for instance, the progression of neonatally acquired infection (BARON et al. 1976; HAYASHI et al. 1983).

6.1.2 B Cell Suppression Suggests that Antibody Reduces the Spread of Virus in the PNS

The previous section alluded to the failure of systemically administered antibody to arrest the progression of established ganglionic infection, from which it might be concluded that antibody has no significant role in regulating the spread of virus within the PNS during primary infection. However, the failure of passively administered antisera to control neurological disease may be the result of an intact blood-brain barrier. MCKENDALL (1983) showed that systemic antibody has limited access to the nervous system until infection is at an advanced stage, by which time high titers of virus may be present. Therefore, a closer look at the role of humoral immunity is warranted.

Based on the presence of Ig-bearing cells in spinal ganglia, it seems that antibody can be synthesized in the immediate vicinity of infected neurons as early as 4 days after inoculation (COOK and STEVENS 1983). The B cell suppressed mouse is a powerful immunological tool (LAWTON et al. 1972) that has been used to determine whether endogenously produced antibody contributes to the control of primary herpes simplex. B cell suppressed BALB/c mice do not produce antibodies in response to immunological challenge with HSV but are otherwise immunologically competent (KAPOOR et al. 1982). Although elimination of the antibody response has no effect on healing of cutaneous herpes, B cell suppressed mice experience a more florid primary infection of the PNS than their fully immunocompetent littermates (SIMMONS and NASH 1987). In the absence of an antibody response, infection is quantitatively greater and more widespread (Fig. 3) and hence a greater number of neurons are presumably exposed to virus. Despite this more florid infection, clearance of virus from the PNS of B cell suppressed mice is not prolonged and no increase in mortality has been observed. It must be concluded that antibody has a subtle and potentially important role in reducing the number of neurons exposed to virus but other mechanisms are

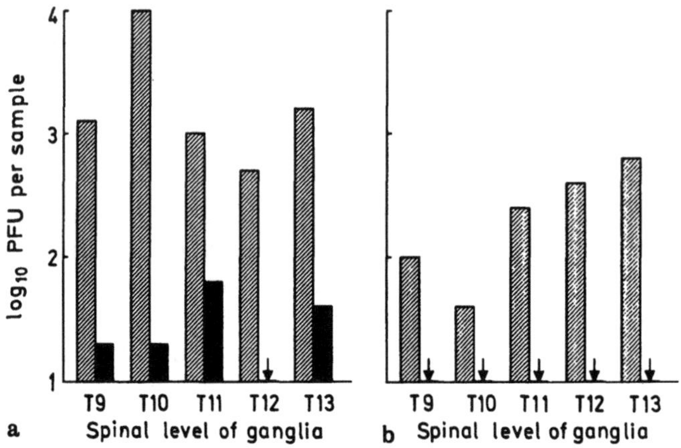

Fig. 3 a, b. B cell suppression. Recovery of virus from spinal ganglia (T9–T13) of normal (*stippled*) and B cell suppressed (*hatched*) BALB/c mice ipsilateral (**a**) and contralateral (**b**) to the inoculation site 8 days after infection with 2×10^5 pfu of HSV-1, strain *SC16*. Mean values from five mice are presented

responsible for removal of infectious material. Clinical observations strongly support this view: for instance, herpes simplex heals normally in children suffering from Bruton's agammaglobulinemia (MERIGAN and STEVENS 1971).

Extensive spread of virus within a ganglion before infection is brought under control has the potential to disseminate latency throughout the neurodermatome. Recrudescence of herpes simplex in humans, though generally confined to a single neurodermatome, often occurs adjacent to the site of primary infection (e.g., gingivostomatitis recurs as perioral cold sores), suggesting that latent infection is indeed established in neurons that do not directly innervate the area originally infected. In the experiments of KAPOOR et al. (1982) latent virus could be reactivated more frequently from the ganglia of mice that had been B Cell suppressed during primary infection than from the ganglia of normal animals, implying that elimination of the humoral response leads to an increase in the number of latently infected neurons. Perhaps the quality and rapidity of the antibody response are factors that determine (among others) how far and wide the virus spreads.

6.1.3 Potential Mechanisms of Antibody Action in the PNS

The mechanism by which antibody dampens the spread of HSV in the PNS is unknown, but consideration of how virus might spread from one primary sensory neuron to another may help to formulate hypotheses. Syncytium formation is a prominent feature in HSV infected cell cultures and cutaneous lesions; direct passage of virus from cell to cell by fusion of membranes is a powerful strategy for the spread of infection without exposing virions to extracellular spaces. With the exception of high titer monoclonal antibodies directed against glycoproteins that mediate membrane fusion (GOMPELS and MINSON 1986), antiserum is generally unable to affect cell-to-cell spread of HSV in cell monolayers (HOOKS et al. 1976) and the skin (SIMMONS and NASH 1987). A striking feature of ganglionic infection is the absence of cell fusion, implying that virus is reliant on an extracellular route to move between one neuron and the next. The soma of primary sensory neurons are never in direct contact and the interspersed capsular cells do not support the full replicative cycle of HSV (COOK and STEVENS 1973; HILL and FIELD 1973). Given the seemingly impervious nature of the capsule that surrounds the body of each neuron it is possible that virus moves from cell to cell outside the confines of the ganglion. The central processes of the neurons communicate within the spinal cord or brainstem and TULLO et al. (1982) have provided direct evidence for a "back door" route of viral spread after infection of the mouse cornea. It is also possible, akin to the round trip hypothesis of KLEIN (1976), that virus emerges from the peripheral processes of neurons into the skin during the early acute infection and there gains access to fresh nerve endings and hence fresh neurons in which to replicate. Passively administered neutralizing (but not nonneutralizing) antibody is able to interrupt the egress of virus at this site, but levels higher than those naturally achieved are required (SIMMONS and NASH 1987).

Based on the above considerations it is tempting to postulate that locally synthesized antibody has the capacity to reduce the spread of infection by neutralizing extracellular virus. This is supported by the observation that protection against viral challenge is readily achieved by immunization with envelope glycoproteins that are major neutralizing targets (see the Chapter by BURKE, this volume). Of the seven (or more) virally encoded glycoproteins associated with the virus envelope, glycoprotein D (gD) is the dominant target for naturally occurring neutralizing antibodies (CRANAGE et al. 1983) in humans with established disease, although antibody to a nonglycosylated capsid protein of approximately 40 kDa dominates the early response to primary infection (EBERLE et al. 1984, 1985).

6.2 Establishment and Maintenance of Latency

Latent infection is observed in agammaglobulinemic humans and antibody deficient experimental animals (SEKIZAWA et al. 1980; KAPOOR et al. 1982), indicating that an antibody response is not a prerequisite for persistence of the viral genome. Nevertheless, several unexplained observations suggest that a role for antibody in discouraging productive infection in favor of latency should not be discounted without careful thought. For instance, PRICE and SCHMITZ (1979) found a higher incidence of latency in the superior cervical (sympathetic) ganglia of mice protected with antiserum than in control animals, and OAKES and LAUSCH (1984) described suppression of replication of HSV in organ cultures of trigeminal ganglia by the continued presence of nonneutralizing monoclonal antibodies directed against gB and gE. The mechanism(s) by which these and other monoclonal antibodies that are inactive in virus neutralization, complement-mediated lysis, and antibody-dependent cellular cytotoxicity assays (RECTOR et al. 1984) mediate an antiviral effect remains obscure.

The finding that virus could be recovered from latently infected ganglia after a period of culture in vitro (STEVENS and COOK 1971) led to the proposal that immune mechanisms were important for maintaining latency in vivo. Transplantation of latently infected ganglia, in Millipore chambers, into either immune or naive recipients showed a relatively poor rate of reactivation in the immune group (STEVENS and COOK 1974). Furthermore, passive transfer of antiserum appeared to reduce reactivation. However, the ability of antibody deficient hosts to stably maintain a latent infection and our current understanding of the molecular nature of latency both strongly refute the possibility that antibody keeps the virus dormant. Early observations presumably reflect a reduction in extracellular spread of virus in ganglia once reactivation has occurred.

7 T Cells

7.1 Control of Primary Infection

It is well established that profound abnormalities of cellular immunity compromise the ability of humans and experimental animals to control herpes simplex virus infection (see the Chapter by SCHMID and ROUSE, this volume). Immunosuppressed humans steer a precarious course through attacks of herpes, between the clinical extremes of chronic, aggressive, cutaneous lesions and disseminated visceral disease. The outcome is probably determined as much by the degree of innate nonadaptive resistance (see Sect. 5.2) as by the extent of specific immunosuppression. However, most inbred mouse strains have poor or mediocre innate resistance to HSV and develop florid neurological infection when immunocompromised. Consequently, murine models, in which T cells can be readily manipulated, provide excellent frameworks in which to develop concepts relating to the adaptive mechanisms that control infection of the PNS. At the same time, basic ground rules for the behavior of T cells have been established by studying them in vitro.

7.1.1 Functional Analysis of T Cells In Vitro

Polyclonal and monoclonal anti-HSV lymphocyte populations have been studied extensively in vitro in the search for effector mechanisms that might be responsible for terminating the replication and spread of HSV in the host. Two mechanistic categories emerged from early studies: (1) direct MHC-restricted cytotoxicity (PFIZENMAIER et al. 1977) and (2) antigen driven secretion of cytokines that might act directly on infected cells or activate macrophages in the vicinity of the infection, as occurs in the delayed-type hypersensitivity (DTH) response (NASH et al. 1980a). While in vitro studies cannot tell us whether cytotoxicity and/or cytokine production are involved in clearance of virus from the nervous system, several relevant themes have emerged as data have been gathered. First, a high proportion of in vitro CTLs recognize HSV immediate early gene products, presumably in the context of class I MHC glycoproteins (MARTIN et al. 1988, 1990). Second, cloned herpes-specific T cells may exhibit a multitude of properties in vitro irrespective of their phenotype/MHC restriction (YASUKAWA and ZARLING 1984; JOHNSON et al. 1990). Finally, different inbred strains of mice and by extrapolation different outbred hosts, recognize a different array of viral antigens as a result of MHC diversity (MARTIN et al. 1990). Therefore conclusions about target recognition apply only to the experimental system used: generalizations are misleading unless proven otherwise.

7.1.2 Adoptive Transfer Experiments

Cells taken from the spleen or draining lymph nodes of HSV infected mice exhibit a wide array of potential antiviral responses (NASH et al. 1980a, b) and, when transferred in sufficient numbers to syngeneic naive animals, they may confer protection against lethal challenge (LARSEN et al. 1983) and/or greatly accelerate virus clearance from the skin (NASH and ASHFORD 1982; SIMMONS and NASH 1984). Adoptive transfer generally prevents the establishment of acute ganglionic infection, but the efficacy of immune cells is lost if their administration is delayed until ganglionic infection is established; under these circumstances recovery proceeds at a normal rather than an accelerated rate.

The protective cell population is comprised of T lymphocytes, and MHC restriction of the antiviral response suggests that at least two types of T cell are involved (HOWES et al. 1979; NASH et al. 1981). This is consistent with experiments showing that $CD4^+$ (NAGAFUCHI et al. 1982; NASH and GELL 1983) and $CD8^+$ (LARSEN et al. 1983; BONNEAU and JENNINGS 1989) enriched populations are individually capable of transferring antiviral immunity to naive mice, although there is some conflict in the literature surrounding this issue, perhaps related to the common use of immunocompetent recipients whose "occupied" lymphoid tissues might prevent normal trafficking of donor cells thus diminishing their effect. The multifunctional nature of lymphoid cells makes it impossible to determine with confidence the mechanism by which donor T cells exert their antiviral effect in vivo, although cooperation with resident lymphocytes, including B cells, has been excluded in some systems by prior immunosuppression (by drugs or irradiation) of recipients, which also creates space for incoming cells (OAKES 1975; SETHI et al. 1983; SIMMONS, unpublished). Full protection of immunocompromised recipients typically requires at least 10^6 cells (NAGAFUCHI et al. 1982), which could presumably act either directly or in concert with resident radiation resistant cells such as macrophages or microglia.

What, then has been the value of adoptive transfer experiments? Certainly they have highlighted the difference between prophylaxis and cytoimmunotherapy. The ability of $CD4^+$ and/or $CD8^+$ donor lymphoid cells to prevent ganglionic infection in syngeneic recipients contrasts starkly with their failure to control established disease. Prevention of ganglionic infection might be secondary to accelerated clearance of virus from the skin, which is maximal when class I and class II MHC loci are matched (NASH et al. 1981), because immunoprophylaxis is not accompanied by an inflammatory response in or around neural tissue (SIMMONS, unpublished). The antithesis of cytoimmunotherapy, namely, specific depletion of lymphocytes ("immunosurgery"), strongly implicates adaptive immunity in the control of herpes, yet the efficacy of adoptive transfer declines rapidly over the 24–48 h period following infection. The reason for this decline is unknown, though we have demonstrated that the zosteriform component of primary cutaneous lesions is not infiltrated by mononuclear cells until substantial damage to the epidermis has occurred, by which time virus replication has peaked (SIMMONS and NASH 1984). Further, we speculate that "shuttling" of virus between the epidermis and PNS is an important aspect of the development of primary lesions and that systemically administered immune cells, like antibodies, do not readily penetrate this privileged environment. We must therefore devise other ways to investigate the neuroimmunological response to HSV.

7.1.3 Evidence that Murine CD8$^+$ Cells Mediate Clearance of Infectious Material from the PNS

A commonly used strategy for studying the role of adaptive immunity in recovery from infection is immunosuppression, and the use of monoclonal antibodies to selectively deplete T cell subsets in vivo is a refinement of this approach (COBBOLD et al. 1984). We targeted CD8$^+$ lymphocytes for further investigation because our previous results suggested that class I loci have a substantial influence on recovery from ganglionic infection (SIMMONS 1989). Mice depleted of CD8$^+$ (Lyt2$^+$) cells retain the ability to mount normal antibody and DTH responses, but have been shown to lose all detectable in vitro cytotoxic activity to a variety of antigens including HSV (COBBOLD et al. 1984; NASH et al. 1987).

In view of the potential for continual interchange of infectious material between the skin and sensory neurons, we examined cutaneous and ganglionic infection in parallel using the flank zosteriform model. The extent to which CD8-selective immunosurgery compromised the ability of experimental animals to resolve a cutaneous infection, judged by clearance of infectious material from the site of inoculation, was mouse strain or more particularly H-2 haplotype, dependent. For the first 5 days of infection, the amount of virus recovered from the skin of normal and CD8-depleted mice was similar, which is consistent with observations made in totally immunodeficient animals and reflects the time taken for an effective primary immune response to develop. Thereafter, it was apparent that CD8 depletion reduced the ability of BALB/c (H-2K^dD^d) and C57B10 (H-2K^bD^b) mice to limit the infection (Fig. 4), suggesting that this subset of lymphocytes may participate in the control of cutaneous herpes. In contrast, despite adequate CD8 depletion (assessed immunohistochemically), CBA (H-

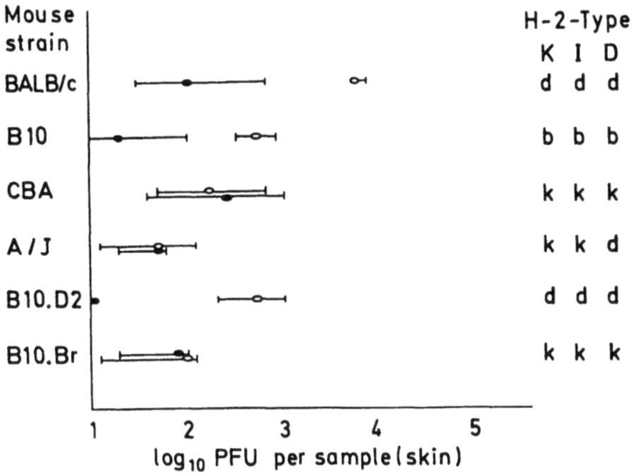

Fig. 4. CD8-selective immunosurgery. Recovery of virus from the skin of various CD8-depleted (*open points*) or immunocompetent (*solid points*) inbred mice 7 days after infection with 2×10^5 pfu of HSV-1, strain *SC16*. CD8-selective immunosurgery was performed using monoclonal antibody YTS169.4 according to previously described protocols (COBBOLD et al. 1984). *Points* represent the geometric mean titer obtained from groups of 5–10 animals; *bars* indicate the range of values

$2K^kD^k$) and A/J (H-$2K^kD^d$) mice were not similarly compromised, strongly implying that $CD8^+$ cells, which include CTLs, make little contribution to HSV clearance from the skin of H-2^k mice. The association of this deficit with H-2^k alleles was strengthened by examining the effect of CD8 depletion on virus clearance from the skin of C57B10 H-2 congenic mice (Fig. 4). While C57B10.D2 (H-2^d) were compromised following immunosurgery, C57B10.Br (H-2^k) were unaffected. The behavior of A/J (H-$2K^kD^d$) animals suggests that the *K* locus alone can have a major influence on the function of herpes-specific $CD8^+$ cells in the skin. This supports a previous in vitro study showing that class I-restricted CTLs, generated in H-2^b mice, preferentially recognize HSV antigens in association with H-$2K^b$ (JENNINGS et al. 1984), and extends the observation that antiviral immunity conveyed by adoptive lymphocyte transfer requires donor-recipient compatibility at H-2*K* (and *I*), but not necessarily at H-2*D* (NASH et al. 1981).

In our system, clearance of infectious virus from spinal ganglia was impaired by CD8-selective immunosurgery in all mice tested, including H-2^k strains. The contribution made by $CD8^+$ cells to the control of infection was highlighted by studying the distribution of viral antigens throughout the PNS 8 days after infection (Fig. 5), at which time immunocompetent animals showed only a small residuum of infection in neurons situated in ganglia at or close to T10. In contrast, in CD8-depleted mice the infection was florid and distributed throughout all the thoracic and lumbar ganglia tested (T6-L1). CD8 depletion impaired the removal of infectious material from dorsal root ganglia as effectively as total T cell depletion, implying that $CD8^+$ cells make a substantial contribution to the overall adaptive immune response to virally infected neurons (SIMMONS, unpublished).

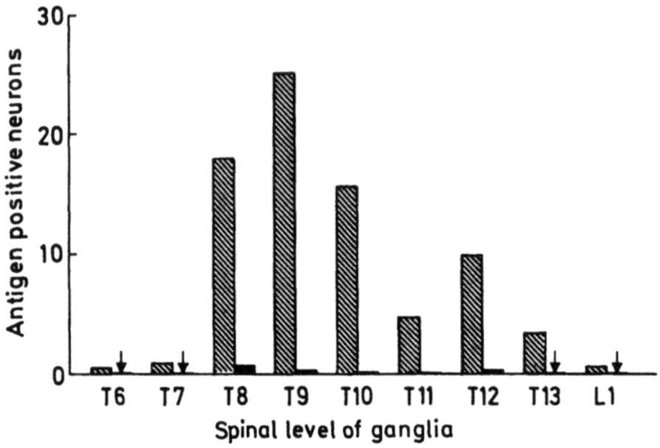

Fig. 5. CD8-selective immunosurgery. Recovery of virus from spinal ganglia (T6-L1) of CD8-depleted (*hatched bars*) and immunocompetent (*stippled bars*) mice 8 days after infection of BALB/c mice with 2×10^5 pfu of HSV-1, strain *SC16*

7.1.4 Fate of Infected Neurons

On exposure to virus, some neurons immediately enter a latent phase of infection (reviewed by STEVENS 1989) whereas other neurons become productively infected. Although virus replication invariably leads to the destruction of HSV infected cells in vitro (reviewed by ROIZMAN and SEARS 1987), recent results from our laboratory (SIMMONS and TSCHARKE 1992) suggest that this is not necessarily the fate of infected neurons in vivo (Table 1). During the entire course of zosteriform spread in the tenth thoracic neurodermatome we found that very few primary sensory neurons were killed, even though at least 13% became infected, as

Table 1. Fate of HSV infected neurons in immunocompetent and CD8-depleted BALB/c mice

	Number of neurons per ganglionic section ± SEM[a]			
	Before infection	Peak[b]	After Infection	
			Immunocompetent	CD8-depleted
Viable neurons[c]	139.4 ± 1.6	n/a	134.7 ± 1.2[d]	88.8 ± 3.3
Antigen-positive	0	18.7 ± 0.7	<1	15.5
VP16-positive	0	12.8 ± 1.6	n/a	n/a

[a] In all cases >300 ganglionic sections from the tenth thoracic neurodermatome were examined in order to determine the relative number of neurons present before and after infection with 2×10^5 pfu HSV-1, strain SC16; n/a, not assessed

[b] A "snapshot" view of ganglia 5 days after inoculation of the flank, giving an estimate of the lower limit of the number of neurons infected during acute infection

[c] Assessed using hematoxylin/eosin and luxol fast blue/cresyl violet staining

[d] In a separate immunocompetent control group for CD8-depleted animals there were 138.8 ± 2.8 neurons/ganglionic section

determined by a "snapshot" immunohistochemical view of ganglia at the peak of infection. The number of neurons (>9%) expressing VP16 (encoded by a late gene), the yield of infectious virus from ganglia in this system (SIMMONS and NASH 1984), and previous electron microscopic studies (DILLARD et al. 1972) all suggest that many neurons survive advanced stages of HSV replication, most likely including the manufacture of infectious virus particles. This reflects a specialized evolutionary relationship between HSV and (1) sensory neurons and (2) the host as a whole, because the CD8$^+$ T cell response is essential for neuronal survival. Depletion of CD8$^+$ cells not only leads to uncontrolled spread of infection (Fig. 5) but also to widespread neuronal death (Table 1). It must be stressed that if mice are overwhelmed by the infection prior to development of CD8$^+$ effector cell activity (detectable 5 days after infection) by, for instance, inoculation of too much virus, the effect is similar to CD8 depletion, i.e., death of neurons during primary infection is prominent. This is a feature of many animal models of herpes simplex as they are commonly used.

There are three implications of the above data. First, the mechanism by which virus replication is terminated does not lyse the infected cell. Second, replication of HSV in neurons in vivo is not invariably cytolytic, representing a highly evolved virus-host relationship. Third, rescued neurons may have the capacity to harbor nonreplicating viral genomes for long periods and as such are a potential source for the bulk of HSV DNA found in the ganglia of humans and experimental animals during latency.

7.1.5 Potential Mechanisms of Action of CD8$^+$ Cells in the PNS

The effect of immunosurgery directed against CD8$^+$ lymphocytes suggests that this population of cells, which includes classic CTLs, plays a pivotal role in terminating HSV replication in the PNS. The mechanism of action is unknown though direct cytotoxicity is excluded on two counts. First, our data show that the majority of infected cells are not killed. Second, neurons are unusual in that they do not express significant amounts of the MHC class I gene products necessary for recognition of antigen by CD8$^+$ CTLs, even when stimulated by IFN-γ (WONG et al. 1984) or infected with HSV (WEINSTEIN et al. 1990). Therefore, it is likely that their beneficial effect is mediated by the release of cytokines within the nervous system. This is supported by many examples of CD8$^+$ T cell clones that can secrete lymphokines in an antigen-specific manner in vitro (see Sect. 7.1.1).

Whatever mechanism underlies the activity of these cells, they need to be stimulated at the site of infection by viral antigen presented in the context of class I MHC and, because neurons cannot fulfill this function, we must look for a surrogate antigen-presenting cell. In spinal ganglia the capsular cells that surround each neuron are counterparts of the microglia of the CNS. During acute infection with HSV, these cells, although not themselves productively infected, strongly express both class I and class II MHC glycoproteins (WEINSTEIN et al. 1990). We therefore hypothesize that capsular or microglial cells are of key importance in activating intraneural antiviral responses and this is a focus of future work.

Landmark observations by OLDSTONE et al. (1986) showed that lymphocytic choriomeningitis virus (LCM) is cleared from the brain of chronically infected mice after adoptive transfer of antiviral $CD8^+$ T cells. There are several major differences and some similarities between our system and the LCM model. The widespread cell death typically associated with removal of LCM from viscera is not apparent in the CNS, emphasizing the potential for cytokine-mediated nonlytic control of viral infection in neurons. However, LCM is cleared slowly, without cellular infiltration of the brain parenchyma, whereas a florid ganglionitis (COOK and STEVENS 1973) that includes a T cell infiltrate (GEBHARDT and HILL 1990) accompanies abrupt termination of HSV replication in the PNS (Fig. 6). Perhaps the most significant difference between LCM and HSV is the ability of the latter to avoid complete clearance from the nervous system by persisting in a nonreplicating state that is apparently invisible to the immune system.

Class I MHC-restricted cytotoxicity cannot be excluded as a contributory mechanism in clearance of virus from the skin. The ability of $CD8^+$ T cells to promote recovery at this site, assessed by selective immunosurgery, is substantially reduced in the context of $H\text{-}2^k$-MHC genes, suggesting a broad hyporesponsiveness of this haplotype to important protective HSV antigens, in the same way that certain mouse H-2 haplotypes are associated with a deficient CTL response to several whole influenza virus proteins (BENNINK and YEWDELL 1988).

We do not know why the H-2 related differences associated with CD8 activity in the skin are not reflected in the PNS, though we believe that the potentially different modes of action at these sites may involve a different array of target viral antigens. For cytotoxicity to be maximally effective as a strategy for controlling infection it makes sense for the immune system to recognize early stages of the viral replicative cycle; indeed, a proportion of anti-HSV CTL precursors recognize immediate early viral gene products. Owing to their inability to synthe-

Fig. 6. Clearance of viral antigens from spinal ganglia of C57B10 mice infected with 2×10^4 pfu of HSV-1, strain *SC16* and killed at 8 hourly intervals

size MHC glycoproteins, neurons do not have the capacity to offer immediate early antigens as targets for direct cytotoxicity. Consequently, in neural tissue there is no selective advantage in responding to a highly restricted array of viral antigens, which is perhaps the reason that the H-2^k related hyporesponsiveness associated with virus clearance from the skin is not manifested in the PNS.

7.2 Control of Reactivation

Many of the issues relating to control of productive infection during the primary phase of herpes simplex may be applicable to virus replication after reactivation. In humans some attention has been paid to factors that determine the likelihood of recrudescence following reactivation events. Subtle depression of lymphokine production, secretion (by T cells) of substances that inhibit cytokine activity (SHERIDAN et al. 1987), and an increase in absolute numbers of suppressor T cells (CAUDA et al. 1989) have all been associated with recrudescence. Lymphocyte proliferation in response to HSV antigens is decreased during the early stages of recrudescence owing to suppressor cell activity (VESTEY et al. 1989) and then rises to above resting levels over the ensuing 2–3 weeks before returning to original values. There is evidence that the magnitude of the T cell response to reactivated herpes, assessed by IFN-γ production, is directly related to the periodicity of recrudescence (CUNNINGHAM and MERIGAN; TORSETH et al. 1987), again suggesting that poor cytokine production might tip the balance towards recrudescence and away from asymptomatic viral shedding.

The preceding observations imply that an antigen driven host response is involved in the control of reactivation in humans, but they do not indicate whether such a response acts primarily in the nervous system or in the skin. Recourse to animal models may be required to gain a deeper understanding of the neuroimmunological events that follow reactivation of HSV in neurons.

8 Concluding Remarks

Despite the irreplaceable nature of neurons, the vertebrate nervous system is a favored site of virus persistence. The ability of neurons to tolerate replication of HSV in the presence of $CD8^+$ lymphocytes is a dramatic illustration of a highly evolved virus-host relationship. We speculate that the ability of the host to terminate productive infection of neurons without killing these vital cells may be a significant selective force in the evolution of viral latency.

Acknowledgments. The authors thank Mrs. C. Bayley for assisting with the preparation of this manuscript. Much of the work presented was supported by project grants 87/0742, 88/0021, and 90/0176 from the National Health and Medical Research Council of Australia.

References

Bak IJ, Markham CH, Cook ML, Stevens JG (1977) Intraaxonal transport of herpes simplex virus in the rat central nervous system. Brain Res 136: 415–429
Bak IJ, Markham CH, Cook ML, Stevens JG (1978) Ultrastructural and immunoperoxidase study of striatonigral neurons by means of retrograde axonal transport of herpes simplex virus. Brain Res 143: 361–368
Baringer JR (1975) Herpes simplex virus infection of nervous tissue in animals and man. Prog Med Virol 20: 1–26
Baron S, Worthington MG, Williams J, Gaines JW (1976) Post exposure serum prophylaxis of neonatal herpes simplex virus infection of mice. Nature 261: 505–506
Bennink JR, Yewdell JW (1988) Murine cytotoxic T lymphocyte recognition of individual influenza virus proteins. High frequency of nonresponder MHC class I alleles. J Exp Med 168: 1935–1939
Blyth WA, Harbour DA, Hill TJ (1984) Pathogenesis of zosteriform spread of herpes simplex virus in the mouse. J Gen Virol 65: 1477–1486
Bonneau RH, Jennings SR (1989) Modulation of acute and latent herpes simplex virus infection in C57BL/6 mice by adoptive transfer of immune lymphocytes with cytolytic activity. J Virol 63: 1480–1484
Cabrera CV, Wohlenberg C, Openshaw H, Ray-Mendez M, Puga A, Notkins AL (1980) Herpes simplex virus DNA sequences in the CNS of latently infected mice. Nature 288: 288–290
Carton CA, Kilbourne ED (1952) Activation of latent herpes simplex by trigeminal sensory root section. N Engl J Med 246: 172–176
Cauda R, Laghi V, Tumbarello M, Ortona L, Whitley RJ (1989) Immunological alternations associated with recurrent herpes simplex genitalis. Clin Immunol Immunopathol 51: 294–302
Cauna N (1973) The free penicillate nerve endings of the human hairy skin. J Anat 115: 277–288
Chan WL, Tizard ML, Faulkner L (1989) Proliferative T-cell response to glycoprotein B of the human herpes viruses: the influence of MHC and sequence of infection on the pattern of cross-reactivity. Immunology 68: 96–101
Cobbold SP, Jayasuriya A, Nash A, Prospero TD, Waldmann H (1984) Therapy with monoclonal antibodies by elimination of T-cell subsets in vivo. Nature 312: 548–551
Cook ML, Stevens JG (1973) Pathogenesis of herpetic neuritis and ganglionitis in mice: evidence for intra-axonal transport of infection. Infect Immun 7: 272–288
Cook ML, Stevens JG (1983) Restricted replication of herpes simplex virus in spinal ganglia of resistant mice is accompanied by an early infiltration of immunoglobulin G-bearing cells. Infect Immun 40: 752–758
Cranage MP, McLean CS, Buckmaster EA, Minson AC, Wildy P, Coombs RRA (1983) The use of monoclonal antibodies in reverse passive haemagglutination tests for herpes simplex antigens and antibodies. J Med Virol 11: 295–306
Cunningham AL (1991) Strategies for control of recurrent herpes simplex. Todays Life Sci 3: 32–40
Cunningham AL, Merigan TC (1983) Gamma interferon production appears to predict time of recurrence of herpes labialis. J Immunol 130: 2397–2400
DeMaeyer-Guignard J, Dandoy F, Bailey DW, DeMaeyer E (1986) Interferon structural genes do not participate in quantitative regulation of interferon production by If loci as shown by C57BL/6 mice that are congenic with BALB/c mice at the alpha interferon gene cluster. J Virol 58: 743–747
Dillard SH, Cheatham WJ, Moses HL (1972) Electron microscopy of zosteriform herpes simplex infection in the mouse. Lab Invest 26: 391–402
Eberle R, Mou S-W, Zaia JA (1984) Polypeptide specificity of the early antibody response following primary and recurrent genital HSV-2 infections. J Gen Virol 65: 1839–1843
Eberle R, Mou S-W, Zaia JA (1985) The immune response to herpes simplex virus: comparison of the specificity and relative titers of serum antibodies directed against viral polypeptides following primary herpes simplex virus type 1 infections. J Med Virol 16: 147–162
Efstathiou S, Minson AC, Field HJ, Anderson JR, Wildy P (1986) Detection of herpes simplex virus-specific DNA sequences in latently infected mice and in humans. J Virol 57: 446–455
Ennis FA (1973) Host defense mechanisms against herpes simplex virus. I. Control of infection in vitro by sensitized spleen cells and antibody. Infect Immun 7: 898–904
Field HJ, Wildy P (1978) The pathogenicity of thymidine kinase-deficient mutants of herpes simplex virus in mice. J Hyg 81: 267–277

Field HJ, Bell SE, Elion GB, Nash AA, Wildy P (1979) Effect of acycloguanosine treatment on acute and latent herpes simplex infections in mice. Antimicrob Ag Chemother 15: 554–561
Foerster O (1933) The dermatomes in man. Brain 56: 1–39
Gallina G, Cumbo V, Messina P, Caprera V, Lio D, Caruso C (1989) Lack of correlation between HLA-B35 resistance against herpes labialis and antibody titers to HSV-1. Oral Surg Oral Med Oral Pathol 68: 167–170
Gebhardt BM, Hill JM (1990) Cellular neuroimmunologic responses to ocular herpes simplex virus infection. J Neuroimmunol 28: 227–236
Gompels U, Minson A (1986) The properties and sequence of glycoprotein H of herpes simplex virus type 1. Virology 153: 230–247
Harbour DA, Hill TJ, Blyth WA (1981) Acute and recurrent herpes simplex in several strains of mice. J Gen Virol 55: 31–40
Harbour DA, Hill TJ, Blyth WA (1983) Recurrent herpes simplex in the mouse: inflammation in the skin and activation of virus in the ganglia following peripheral stimulatio. J Gen Virol 64: 1491–1498
Hayashi Y, Wada T, Mori R (1983) Protection of newborn mice against herpes simplex virus infection by prenatal and postnatal transmission of antibody. J Gen Virol 64: 1007–1012
Hill TJ, Field HJ (1973) The interaction of herpes simplex virus with cultures of peripheral nervous tissue: an electron microscopic study. J Gen Virol 21: 123–133
Hill TJ, Field JH, Blyth WA (1975) Acute and recurrent infection with herpes simplex virus in the mouse: a model for studying latency and recurrent disease. J Gen Virol 28: 341–353
Hooks JJ, Burns W, Hayashi K, Geis S, Notkins AL (1976) Viral spread in the presence of neutralizing antibody: mechanisms of persistence in foamy virus infection. Infect Immun 14: 1172–1178
Howes EL, Taylor W, Mitchison NA, Simpson E (1979) MHC matching shows that at least two T-cell subsets determine resistance to HSV. Nature 277: 67–68
Hsiung GD, Mayo DR, Lucia HL, Landry ML (1984) Genital herpes: pathogenesis and chemotherapy in the guinea pig model. Rev Infect Dis 6: 33–50
Jennings SR, Rice PL, Pan S, Knowles BB, Tevethia SS (1984) Recognition of herpes simplex virus antigens on the surface of mouse cells of the $H\text{-}2^b$ haplotype by virus-specific cytotoxic T lymphocytes. J Immunol 132: 475–481
Johnson RM, Lancki DW, Fitch FW, Spear PG (1990) Herpes simplex virus glycoprotein D is recognized as antigen by $CD4^+$ and $CD8^+$ T lymphocytes from infected mice. Characterization of T cell clones. J Immunol 145: 702–710
Kapoor AK, Nash AA, Wildy P (1982) Pathogenesis of herpes simplex virus in B cell-suppressed mice: the relative roles of cell-mediated and humoral immunity. J Gen Virol 61: 127–131
Katz JP, Bodin ET, Coen DM (1990) Quantitative polymerase chain reaction analysis of herpes simplex virus DNA in ganglia of mice infected with replication-incompetent mutants. J Virol 64: 4288–4295
Klein RJ (1976) Pathogenetic mechanisms of recurrent herpes simplex virus infections. Arch Virol 51: 1–13
Knipe DM (1989) The role of viral and cellular nuclear proteins in herpes simplex virus replication. Adv Virus Res 37: 85–123
Kristensson K, Lycke E, Sjöstrand (1971) Spread of herpes simplex virus in peripheral nerves. Acta Neuropathol (Berl) 17: 44–53
Larsen HS, Russel RG, Rouse BT (1983) Recovery from herpes simplex virus type 1 infection is mediated by cytotoxic T lymphocytes. Infect Immun 41: 197–204
Lawton A, Asofsky R, Hylton M, Cooper M (1972) Suppression of immunoglobulin class synthesis in mice. J Exp Med 135: 277–297
Leib DA, Coen DM, Bogard CL, Hicks KA, Jager DR, Knipe DM, Tyler KL, Schaffer PA (1989) Immediate-early regulatory gene mutants define different stages in the establishment and reactivation of herpes simplex virus latency. J Virol 63: 759–768
Lopez C (1975) Genetics of natural resistance to herpesvirus infections in mice. Nature 258: 152–153
Lopez C (1980) Resistance to HSV-1 in the mouse is governed by two major, independently segregating non H-2 loci. Immunogenetics 11: 87–92
Lycke E, Kristensson K, Svennerholm B, Vahlne A, Ziegler R (1984) Uptake and transport of herpes simplex virus in neurites of rat dorsal root ganglia cells in culture. J Gen Virol 65: 55–64
Marrack P, Kappler J (1987) The T-cell receptor. Science 238: 1073–1079
Martin S, Courtney RJ, Fowler G, Rouse BT (1988) Herpes simplex virus type 1-specific cytotoxic T lymphocytes recognize virus nonstructural proteins. J Virol 62: 2265–2273

Martin S, Zhu XX, Silverstein SJ, Courtney RJ, Yao F, Jenkins FJ, Rouse BT (1990) Murine cytotoxic T lymphocytes specific for herpes simplex virus type 1 recognize the immediate early protein ICP4 but not ICP0. J Gen Virol 71: 2391–2399
McKendall R (1983) Delayed IgG-mediated clearance of herpes simplex virus type 1 from the CNS but not footpad during the early stages of infection: possible result of relative integrity of the blood-brain barrier. J Gen Virol 64: 1965–1972
McKendall RR, Klassen T, Baringer JR (1979) Host defences in herpes simplex infection of the nervous system: effect of antibody on disease and viral spread. Infect Immun 23: 305–311
McLean JL, Darby G (1980) Herpes simplex virus latency: the cellular location of virus in dorsal root ganglia and fate of the infected cell following virus activation. J Gen Virol 51: 233–243
Merigan TC, Stevens DA (1971) Viral infections in man associated with acquired immunological deficiency states. Fed Proc 30: 1858–1864
Mok CH (1971) Zoster-like disease in infants and young children. N Engl J Med 285: 294
Music SI, Fine EM, Togo Y (1971) Zoster-like disease in the newborn due to herpes simplex virus. N Engl J Med 284: 20–24
Nagafuchi S, Hayashida I, Higa K, Wada T, Mori R (1982) Role of Lyt-2 positive immune T cells in recovery from herpes simplex virus infection in mice. Microbiol Immunol 26: 359–362
Nash AA, Ashford NPN (1982) Split T-cell tolerance in herpes simplex virus-infected mice and its implication for antiviral immunity. Immunology 45: 761–767
Nash AA, Gell PGH (1983) Membrane phenotype of murine effector and suppressor T cells involved in delayed hypersensitivity and protective immunity to herpes simplex virus. Cell Immunol 75: 348–355
Nash AA, Field HJ, Quartey-Papafio R (1980a) Cell-mediated immunity in herpes simplex virus-infected mice: induction, characterization and antiviral effects of delayed type hypersensitivity. J Gen Virol 48: 351–357
Nash AA, Quartey-Papafio R, Wildy P (1980b) Cell mediated immunity in herpes simplex virus-infected mice: functional analysis of lymph node cells during periods of acute and latent infection, with reference to cytotoxic and memory cells. J Gen Virol 49: 309–317
Nash AA, Phelan J, Wildy P (1981) Cell-mediated immunity in herpes simplex virus-infected mice: H-2 mapping of the delayed-type hypersensitivity response and the antiviral T cell response. J Immunol 126: 1260–1262
Nash AA, Jayasuriya A, Phelan J, Cobbold SP, Waldmann H, Prospero T (1987) Different roles for L3T4+ and Kyt 2+ T cell subsets in the control of an acute herpes simplex virus infection of the skin and nervous system. J Gen Virol 68: 825–833
Oakes JE (1975) Role for cell mediated immunity in the resistance of mice to subcutaneous herpes simplex virus infection. Infect Immun 12: 166–172
Oakes JE, Lausch RN (1984) Monoclonal antibodies suppress replication of herpes simplex virus type 1 in trigeminal ganglia. J Virol 51: 656–661
Oldstone MBA, Blount P, Southern PJ, Lampert PW (1986) Cytoimmunotherapy for persistent virus infection reveals a unique clearance pattern from the central nervous system. Nature 321: 239–243
Pfizenmaier K, Starzinski-Powitz A, Rollinghoff M, Falke D, Wagner H (1977) T-cell-mediated cytotoxicity against herpes simplex virus-infected target cells. Nature 265: 630–632
Plummer G, Hollingsworth DC, Phuangsab A, Bowling CP (1970) Chronic infection by herpes simplex viruses and by horse and cat herpes viruses. Infect Immun 1: 351–355
Price RW, Schmitz J (1979) Route of infection, systemic host resistance, and integrity of ganglionic axons influence acute and latent herpes simplex virus infection of the superior cervical ganglion. Infect Immun 23: 373–383
Rabin ER, Jenson AB, Melnick JL (1968) Herpes simplex virus in mice: electron microscopy of neural spread. Science 162: 126–127
Rector JT, Lausch RN, Oakes JE (1984) Identification on infected cell-specific monoclonal antibodies and their role in host resistance to ocular herpes simplex virus type 1 infection. J Gen Virol 65: 657–661.
Richards JT, Kern ER, Overall JC, Glasgow LA (1981) Differences in neurovirulence among isolates of herpes simplex virus types 1 and 2 in mice using four routes of inoculation. J Infect Dis 144: 464–471
Rock DL, Fraser NW (1983) Detection of HSV-1 genome in central nervous system of latently infected mice. Nature 302: 523–525
Roizman B, Sears AE (1987) An enquiry into the mechanisms of herpes simplex virus latency. Annu Rev Microbiol 41: 543–571

Russell AS, Schlaut (1977) Association of HLA-AI antigen and susceptibility to cold sores. Arch Dermatol 113: 1721–1722
Scalzo AA, Fitzgerald NA, Simmons A, La Vista AB, Shellam GR (1990) *Cmv-1*, a genetic locus that controls murine cytomegalovirus replication in the spleen. J Exp Med 171: 1469–1483
Sekizawa T, Openshaw H, Wohlenberg C, Notkins AL (1980) Latency of herpes simplex virus in absence of neutralizing antibody: model for reactivation. Science 210: 1026–1028
Sethi KK (1983) Effects of monoclonal antibodies directed against herpes simplex virus-specified glycoproteins on the generation of virus-specific and H-2-restricted cytotoxic T-lymphocytes. J Gen Virol 64: 2033–2037
Sethi KK, Omata Y, Schneweis KE (1983) Protection of mice from fatal herpes simplex virus type 1 infection by adoptive transfer of cloned virus-specific and H-2 restricted cytotoxic T lymphocytes. J Gen Virol 64: 443–447
Severin MJ, White RJ (1968) The neural transmission of herpes simplex virus in mice. Light and electron microscopic findings. Am J Pathol 53: 1009–1020
Sheridan JF, Beck M, Smith CC, Aurelian L (1987) Reactivation of herpes simplex virus is associated with production of a low molecular weight factor that inhibits lymphokine activity in vitro. J Immunol 138: 1234–1239
Simmons A (1989) H-2-linked genes influence the severity of herpes simplex virus infection of the peripheral nervous system. J Exp Med 169: 1503–1507
Simmons A, La Vista AB (1989) Neural infection in mice after cutaneous incoulation with HSV-1 is under complex host genetic control. Virus Res 13: 263–270
Simmons AA, Nash AA (1984) Zosteriform spread of herpes simplex virus as a model of recrudescence and its use to investigate the role of immune cells in prevention of recurrent disease. J Virol 52: 816–821
Simmons A, Nash AA (1985) Role of antibody in primary and recurrent herpes simplex virus infection. J Virol 53: 944–948
Simmons A, Nash AA (1987) Effect of B cell suppression on primary infection and reinfection of mice with herpes simplex virus. J Infect Dis 155: 649–654
Simmons A, Tscharke DC (1992) Anti-CD8 impairs clearence of HSV from the nervous system: implications for the fate of virally infected neurons. J Exp Med (in press)
Speck PG, Simmons A (1991) Divergent molecular pathways of productive and latent infection with a virulent strain of herpes simplex virus type 1. J Virol 65: 4001–4005
Stanberry LR, Kern ER, Richards JT, Abbott TM, Overall JC (1982) Genital herpes in guinea pigs: pathogenesis of the primary infection and description of recurrent disease. J Infect Dis 146: 397–404
Steiner I, Spivack JG, Deshmane SL, Ace CI, Preston CM, Fraser NW (1990) A HSV 1 mutant containing a nontransinducing Vmw65 protein establishes latent infection in vivo in the absence of viral replication and reactivates efficiently from explanted trigeminal ganglia. J Virol 64: 1630–1638
Stevens JG (1989) Human herpesviruses: a consideration of the latent state. Microbiol Rev 53: 318–332
Stevens JG, Cook ML (1971) Latent herpes virus in spinal ganglia of mice. Science 173: 843–845
Stevens JG, Cook ML (1974) Maintenance of latent herpetic infection. An apparent role for anti-viral IgG. J Immunol 113: 1685–1693
Swain SL (1983) T cell subsets and the recognition of MHC class. Immunol Rev 74: 129–142
Tanaka S, Southam CM (1965) Zoster-like lesions from herpes simplex virus in newborn rats. Proc Soc Exp Biol Med 120: 56–59
Teague O, Goodpasture EW (1923) Experimental herpes zoster. J Med Res 24: 185–200
Torseth JW, Nicholoff BJ, Basham TW, Merigan T (1987) Beta interferon produced by keratinocytes in human cutaneous infection with herpes simplex virus. J Infect Dis 155: 641–648
Tullo AB, Shimeld C, Blyth WA, Hill TJ, Easty DL (1982) Spread of virus and distribution of latent infection following ocular herpes simplex in the nonimmune and immune mouse. J Gen Virol 63: 95–100
Ugolini G, Kuypers HGJM, Simmons A (1987) Retrograde transneuronal transfer of herpes simplex virus type 1 (HSV1) from motoneurones. Brain Res 422: 242–256
Vestey JP, Norval M, Howie S, Maingay J, Neill WA (1989) Variation in lymphoproliferative responses during recrudescent orofacial herpes simplex virus infections. Clin Exp Immunol 77: 384–390
Walz MA, Price RW, Notkins AL (1974) Latent ganglionic infection with herpes simplex virus types 1 and 2: viral reactivation in vivo after neurectomy. Science 184: 1185–1187
Weinstein DL, Walker DG, Akiyama H, McGeer PL (1990) Herpes simplex virus type I infection of the

CNS induces major histocompatibility complex antigen expression on rat microglia. J Neurosci Res 26: 55–65
Whitear M (1960) An electron microscope study of the cornea in mice, with special reterence to the innervation. J Anat 94: 387–409
Wildy P (1967) The progression of herpes simplex virus to the central nervous system of the mouse. J Hyg 65: 173–192
Wildy P, Field HJ, Nash AA (1982) Classical herpes latency revisited. Symposium 33, Society for General Microbiology. Cambridge University Press, Cambridge
Wong GHW, Bartlett PF, Clark-Lewis I, Battye F, Schrader JW (1984) Inducible expression of H-2 and Ia antigens on brain cells. Nature 310: 688–691
Worthington M, Conliffe MA, Baron S (1980) Mechanism of recovery from systemic herpes simplex virus infection I. Comparative effectiveness of antibody and reconstitution of immune spleen cells on immunosuppressed mice. J Infect Dis 142: 163–174
Yasukawa M, Zarling JM (1984) Human cytotoxic T-cell clones directed against herpes simplex virus-infected cells. II. Bifunctional clones with cytotoxic and virus-induced proliferative activities exhibit herpes simplex virus types 1 and 2 specific or type common reactivities. J Immunol 133: 2737–2742
Zawatzky R, Gresser I, DeMaeyer E, Kirchner H (1982a) The role of interferon in the resistance of C57BL/6 mice to various doses of herpes simplex virus type 1. J Infect Dis 146: 405–410
Zawatzky R, Kirchner H, DeMaeyer-Guignard J, DeMaeyer E (1982b) An X-linked locus influences the amount of circulating interferon induced in the mouse by herpes simplex virus type 1. J Gen Virol 63: 325–332
Ziegler RJ, Herman RE (1980) Peripheral infection in culture of rat sensory neurons by herpes simplex virus. Infect Immun 28: 620–623

The Role of T Cell Immunity in Control of Herpes Simplex Virus

D. S. SCHMID[1] and B. T. ROUSE[2]

1 Introduction

Since their characterization in the mid-1960s, thymus-derived, or T lymphocytes have been shown to participate in virtually every process of mammalian immunity. This participation can take the form of response modification, mediated by secreted physiologically active cytokines, or direct effector activity, as exemplified by cell-mediated cytolysis and the inhibition of viral replication by interferon-γ (IFN-γ).

T lymphocytes are superficially indistinguishable from B lymphocytes. The specific antigen receptors for both cell types reflect a common ancestry both in structure and in the means by which they diversify, yet the two cells recognize nominal antigen in fundamentally different ways. B lymphocytes ordinarily recognize antigens in their native state, whereas T lymphocytes can only recognize processed antigen that has bound into a cleft located on the NH_2-terminal portion of a major histocompatibility complex (MHC) molecule (TOWNSEND et al. 1985, 1986; TAYLOR et al. 1987; BODMER et al. 1989; MEUER et al. 1984). In recent

[1] Viral Exanthems and Herpesvirus Branch, 1600 Clifton Rd., G18/7/206, Atlanta, GA 30333, USA
[2] Department of Microbiology, College of Veterinary Medicine, University of Tennessee, Knoxville, TN 37996-0845, USA

Current Topics in Microbiology and Immunology, Vol. 179

years it has been determined that this peptide-MHC antigen moiety is recognized by a single receptor molecule complexed with CD3 on the surface of T lymphocytes (RUDD et al. 1989).

Thus, the receptor molecules of T cells and B cells reflect the different strategies of the two sublineages. B cells see native antigen via their surface immunoglobulin receptor and respond by secreting a soluble form of the receptor molecule itself. T lymphocytes detect alterations in self, as evidenced by the appearance of peptides on MHC which will interact at high affinity with one or more T cell receptor idiotypes in the selected repertoire of an individual. This ability to recognize viral peptides on infected cells enables T cells to provide a central role in the elimination of most viral threats.

A wide variety of immune mechanisms have been reported to act on herpesviruses, including cytotoxic T cells (ROUSE 1984a), natural killer (NK) cells (LOPEZ 1984), humoral immunity (SHORE and NAHMIAS 1981), antibody-dependent cellular cytotoxicity (KOHL 1984), mononuclear phagocytes (MORAHAN 1984), and cytokines (ROUSE 1984b). Since there are so many activities potentially involved in the control of herpesvirus disease, it is often difficult to draw unambiguous conclusions regarding the relative contribution of a single mechanism. This is further complicated by the fact that the evidence implicating, for example, T cell-mediated immunity in HSV infections is largely circumstantial, derived as it is from in vitro human studies and various animal models. Some caution must be taken in the interpretation of in vitro human studies, because they may not accurately reflect in vivo events, and similar care must be exercised in the consideration of results obtained from animal models, most of which present at best a poor analogy to human disease. Nevertheless, the ease with which herpes-specific cytotoxic T lymphocyte (CTL) activity can be demonstrated in a wide variety of systems, the evidence from clinical observations in humans, and the ability of adoptively transferred CTLs to protect in animal models all argue persuasively for a prominent role for CTLs in the control of herpes simplex virus (HSV).

2 Human Studies

2.1 Clinical Observations

Only a minority of individuals present with clinical symptoms during primary infection with HSV and, of these, most develop relatively mild and localized vesicular lesions. Severe and/or disseminated HSV disease is comparatively rare and for the most part restricted to individuals who are in some fashion immunocompromised. Observations of this sort led quite naturally to the conclusion that the control of HSV disease, including periodic recrudescence, is a function of the immune system. Beyond the broad observation that immune mechanisms contribute to the control of HSV disease, it is difficult to sort out the extent to which individual mechanisms contribute to this process on the basis of clinical studies. Furthermore, it is clear that the immune system is not the only factor exerting control over HSV infections, since only a small percentage of immunosuppressed patients ever develop severe HSV disease (SHORE and NAHMIAS 1981).

One inferential piece of evidence that cellular mechanism play a predominant role in curtailing recurrent HSV lesions derives from the apparent failure of circulating antibody to have any effect on recrudescent disease. The level of serum neutralizing antibody specific for HSV does not change significantly with the occurrence of lesions (DOUGLAS and COUCH 1970). In addition, recurrent lesions develop in HSV-seropositive individuals despite very impressive titers of anti-HSV antibody (ZWEERINK and STANTON 1981). In addition, disseminated HSV disease is not observed in patients with selective deficiencies in humoral immunity (SHORE and NAHMIAS 1981).

Patients with depressed T cell function may develop severe HSV disease, whereas persons with immunoglobulin defects appear competent to control HSV infections (WHITLEY 1985; MERIGAN and STEVENS 1971; COREY and SPEAR 1986). Adults infected with HIV-1, which targets and kills $CD4^+$ T cells, frequently develop severe HSV lesions (QUINNAN et al. 1984; SIEGAL et al. 1981), an observation of potential importance since a large proportion of the human CTL response to HSV appears to be mediated by T cells of that phenotype (SCHMID 1988). These cases are also difficult to interpret because the depletion of $CD4^+$ cells has far-reaching effects on the competency of the entire immune system, not just on CTL activity.

Stromal keratitis, a leading cause of blindness in developed countries, appears to result from the induction of an excessive T cell response to HSV, a phenomenon which is discussed in greater detail in another chapter in this volume. This suggests that a delicate balance between what constitutes an effective vs a destructive immune response is tripped in favor of the latter in patients suffering from this disorder.

The immunosuppressive regimen associated with organ transplantation has been commonly reported to result in an increased frequency of recurrent HSV disease (RAND et al. 1976; PASS et al. 1979; HO 1977; SPENCER and ANDERSON 1970). This transplantation-linked effect has been substantially ameliorated with the adoption of cyclosporin A as an immunosuppressive therapy, a treatment which, unlike high dose corticosteroid treatment, does not indiscriminately impair T cell activity.

2.2 In Vitro Studies

Prior to the early 1980s, in vitro examinations of CTL activity in humans were difficult to perform and necessarily limited in scope due to the unavailability of reagents required to identify CTL subsets and owing to the scarcity of suitable target cells. The advent of commercially produced monoclonal subsetting antibodies, together with a simple method of establishing permanent human cell lines by transforming normal B lymphocytes with Epstein-Barr virus, has made the human in vitro CTL system among the most versatile and practical.

One early effort to derive and characterize a number of human CTL clones specific for HSV-1 or HSV-2 led to the surprising observation that all of them where $CD4^+$ rather than $CD8^+$ T cells (YASUKAWA and ZARLING 1984a, b). Most of these clones were restricted in their recognition of antigen to MHC-DR antigens. A study comparing panels of clones generated against HSV-1 or HSV-2 revealed the presence of both type-specific and type-common T cell idiotypes. All clones were also found to be bifunctional, possessing both cytotoxic and helper cell activity (YASUKAWA and ZARLING 1984b). The most popular interpretation of these results has been that long-term culture of lymphocytes with inactivated virus preparations expands a small population of aberrant T cells and that the $CD4^+$ killer cells do not play a significant role in vivo. More recently, however, CTLs stimulated in short-term bulk cultures of peripheral blood lymphocytes with HSV-1 have been shown to consist predominantly of $CD4^+$, $CD8^-$ T lymphocytes (SCHMID 1988; TORPEY et al. 1989). These cells clearly occur in the peripheral blood of seropositive donors at very high frequency, ranging from 1/4000–1/20000, with an average frequency of 1/12000 (SCHMID 1988; SCHMID and MAWLE, unpublished observation). As such, it seems highly unlikely that they represent a physiological anomaly. $CD8^+$ CTLs have also been identified in bulk culture, albeit at lower frequency (TORPEY et al. 1989; CUNNINGHAM and NOBLE 1989; YASUKAWA et al. 1989). The assertion that the mode of antigen presentation skews the human response in vitro does have some validity, however. YASUKAWA et al. (1989) have shown that free HSV antigen preferentially stimulates $CD4^+$ CTL activity, whereas HSV-1-infected fibroblasts preferentially elicit $CD8^+$ CTLs. Nevertheless, under these different conditions of antigen presentation, both $CD4^+$ and $CD8^+$ CTLs are generated. Both the endogenous and exogenous modes of antigen presentation would be expected to occur within a HSV lesion. Thus, even when this factor is taken into account, HSV-specific $CD4^+$ CTLs would probably outnumber $CD8^+$ cells on the basis of these in vitro studies. The central difficulty with accepting a traditional CTL role for $CD4^+$ CTLs in vivo is the apparent rarity of targets which express the necessary class II MHC antigens required by these killer cells in their recognition of nominal antigen. This discrepancy has led some to propose an immunoregulatory role for $CD4^+$ CTLs (BRAAKMAN et al. 1987; OTTENHOFF and MUTIS 1990). Thus, the $CD4^+$ killers are proposed to recognize and kill antigen-presenting cells which carry viral antigens on their surface, by which means the antigen-specific response is dampened. While this almost certainly represents one of the in vivo tasks of the $CD4^+$ CTLs, we propose an alternative hypothesis for the importance of these cells in containing localized viral infections in a separate section of this review.

3 Animal Models

3.1 Murine Studies

Except in a few isolated applications, studies of HSV disease in the mouse do not serve as a true analogue for human disease. However, the availability of a wide variety of MHC syngeneic and recombinant strains of mice, together with a broad selection of lymphocyte subsetting reagents, makes the murine system both a convenient and informative experimental tool for answering basic questions about mammalian immunity to HSV.

3.1.1 In Vitro Analysis of the CTL Response to HSV

The earliest definitive characterizations of HSV-specific, MHC-restricted CTLs were carried out in mice (PFIZENMAIER et al. 1977; SETHI and BRANDIS 1977). In general, the CTL responses against HSV measured in mice have been weak. Viewed historically, the data in these early experiments also frequently included unaccounted for components of nonspecific cytolytic activity. The demonstration of in vivo antiviral CTL induction has been difficult, an obstacle which is by no means limited to the HSV response (ZINKERNAGEL and DOHERTY 1979). One hypothesis is that suppressor cell activity is inhibiting the expression of CTL activity in vivo. It has been shown that the preexposure of mice to cyclophosphamide, a treatment which preferentially impairs suppressor cell activity, with subsequent HSV infection results in an enhancement of the CTL response (LAWMAN et al. 1980b; PFIZENMAIER et al. 1977). In further support of that hypothesis, both antigen-specific and nonspecific suppressor activity have been demonstrated in an in vitro system (HOROHOV et al. 1986). Finally, the rate at which HSV-specific helper T cell precursors develop in limiting dilution has been shown to be affected by $CD8^+$ T cells (PRYMOWISZ et al. 1985). In test wells depleted of $CD8^+$ cells, the frequency of HSV-specific helper T cells increases two to threefold, suggesting a significant component of immunosuppresion.

It has been argued that HSV stimulates a weak response in animals because of a failure to induce adequate levels of T cell help. The basis of this argument comes from an analysis of the precursor frequency by limiting dilution (ROUSE and WAGNER 1984) and is inferred from an earlier study in which helper T cells were reported to have different antigen activation requirements than CTLs (SCHMID and ROUSE 1983). Lymph node cells from HSV-primed mice which had been cultured in the presence of virus alone displayed precursor frequencies that were tenfold lower than those obtained from cells cultured in the presence of both virus and cytokine-rich supernatant. In these experiments, CTL frequencies averaging 1/30 000 were obtained even in the absence of exogenous lymphokines. While that is a typical CTL frequency for an antiviral response, it is also quite possibly below the level of activity that could be detected directly in bulk culture. For example, cloned antiviral CTLs tested at an effector: target cell ratio of 1:1

might, under the best circumstances, result in a specific release value of approximately 20% (KAPLAN et al. 1984; VAN BINNENDIJK et al. 1989). Typically, a single test well in a bulk culture CTL assay would contain 10000 radiolabeled target cells. Conservatively assuming that only 1 in 100 cells in such highly enriched population of antigen-specific killers is active, that leaves approximately 100 CTL to effect what is, by any measure, a marginal release of isotope from the targets. Returning to the freshly isolated cells from an HSV-primed mouse, the number of potential killer cells in a single test well, even at a ratio of 100:1, is only 30 cells. Thus, to some degree at least, the failure to observe evidence of priming for CTLs in directly isolated lymphocytes simply reflects the need to substantially expand a rare population of T cells. Additional support for this hypothesis comes from the observation that HSV-specific CTLs can be detected in splenocytes from mice which have been repeatedly exposed to antigen (SETHI and BRANDIS 1977).

In contrast, HSV-specific memory cells can be readily stimulated with virus in vitro to high levels of activity that is both antigen-specific and MHC-restricted (LAWMAN et al. 1980b; ROUSE and LAWMAN 1980). The induction of a secondary in vitro response to HSV has been shown to require the participation of several cell types or the cytokines they produce (ROUSE and LAWMAN 1980; SCHMID et al. 1981, 1982; SCHMID and ROUSE 1983). The selective removal of adherent cells from primed splenocytes results in a cell population that is refractory to the in vitro stimulation of anti-HSV CTL activity. These adherent cells presumably are required to present antigen to T cell precursors, especially to antigen-specific helper cells, since the antigen-responsiveness of CTLs can be restored by the addition of exogenous helper cell cytokines (ROUSE and LAWMAN 1980; SCHMID et al. 1981). Further support of this hypothesis comes from the observation that the adherent cell population must express the surface class II MHC antigens that T helper cells in general require for their recognition of nominal antigen (SCHMID et al. 1982). A requirement has also been shown in the HSV system for the production of interleukin-1 (IL-1) by the antigen-presenting cell population (SCHMID et al. 1982). Several products of antigen-activated T helper cells have also been demonstrated to be necessary for in vitro CTL responses, including interleukin-2 (IL-2) (ROUSE and LAWMAN 1980; SCHMID et al. 1981), IFN-γ (FARRAR et al. 1981, 1982), and T cell differentiation factor (WAGNER et al. 1982; RAULET and BEVAN 1982), now thought to be interleukin-7 (IL-7) (BERTAGNOLLI and HERRMANN 1990).

One hypothesis which has grown out of this in vitro dissection of the HSV-specific CTL response is that the adherent antigen-presenting cell population is an absolute requirement only for eliciting nonspecific T cell help.

The murine system has been used to identify some of the target antigens for the CTL response to HSV. As with influenza virus (ZWEERINK et al. 1977), the herpes CTL response includes both type-specific and cross-reactive activity (EBERLE et al. 1981). This is hardly surprising considering the extent of genetic relatedness of HSV-1 and HSV-2. It has been estimated that the amount of sequence homology between the two types of HSV is on the order of 46% (KIEFF et al. 1972). Early experimental efforts to define the HSV target antigens for CTLs focused almost exclusively on glycoproteins because it was not yet understood that T cells were capable of recognizing internal antigens. Since T cells recognize only processed peptide in association with an MHC molecule, any protein which is synthesized from a viral genome is a potential target antigen.

Targets infected with a temperature-sensitive mutant deficient in the expression of glycoprotein B (gB) gB and gC were observed to be substantially less susceptible to lysis by herpes-specific CTLs at the nonpermissive temperature (LAWMAN et al. 1980a). The temperature-sensitive mutant used in this study expressed gD at normal levels regardless of temperature. In the same report, infection of targets in the presence of tunicamycin, which prevents the transport of glycoproteins to the cell surface, dramatically reduced the susceptibility of targets to CTL lysis. This led to the conclusion that gC and gB, but not gD, were major target antigens for the HSV CTL response. It should be noted that this early study could not take into account the role of HSV glycoproteins which had not been characterized at the time, including gE, gH, gI, gJ, and the type-specific glycoprotein gG. More recent investigations have taken advantage of steadily evolving recombinant DNA technology as a means for examining the relative contributions of individual HSV-encoded proteins to the CTL response. Lytic activity has been demonstrated against a variety of HSV proteins, the genes for which have been cloned and expressed in an appropriate vector, such as vaccinia, adenovirus-5, or beculovirus. The results of such studies, from several different laboratories, are somewhat difficult to reconcile. For example, gC, but not gB, gD, or gE of HSV-1 could be recognized on target cells by HSV-specific CTLs in one murine study (ROSENTHAL et al. 1987). In another CTLs did not respond to gD but were elicited against gB (BLACKLAWS and NASH 1990). GLORIOSO and coworkers (1985) reported that antigenic determinants on gC were immunodominant in the murine response to HSV. gD has been shown by two groups to activate antigen-specific class II-restricted T cells. Furthermore, the responding cells enhance the efficiency with which test animals clear HSV from local lesions (BLACKLAWS et al. 1987; MARTIN and ROUSE 1987). In neither case was class I-restricted CTL activity in evidence. Yet another study demonstrated that gD could be recognized by both $CD8^+$ and $CD4^+$ CTLs in mice (JOHNSON et al. 1990). When adenovirus-5-vectored glycoproteins (gB, gC, gD, and gE) were used to infect target cells derived from an H-2^k mouse, CTL-mediated lysis could be demonstrated against the gC-infected target (ROSENTHAL et al. 1987). A more recent study by the same group suggested that whichever of the herpes-encoded antigens will serve as targets for CTLs is dependent on the MHC haplotype of the responding animal and on the conditions used to generate specific CTL (WITMER et al. 1990). In those studies, CTL from H-2^k mice could recognize and kill targets transfected with an adenovirus/gC construct but were unable to recognize gB- or gD-positive targets. In contrast, H-2^b and H-2^d mice could recognize gB-positive targets but not gC-positive targets. MESTER et al. (1990) reported on the immunogenicity of a set of overlapping peptides identifying an epitope on gB. This study also revealed strain-specific variation in the immune response but, more importantly, showed that immunization with an appropriately selected peptide could induce a protective response, as measured by the ability to clear an epithelial infection. In addition, one of the synthetic peptides used by this group was a "silent" epitope, in that it was unrecognized by antibodies generated against whole HSV-1 and yet it still generated a protective response when administered to naive animals.

Studies of influenza-specific CTL clones indicate that CTLs arising in two animals with disparate MHC profiles recognize a completely different set of peptides as specific antigens (TAYLOR et al. 1987). The observations in the HSV system suggest that immune responsiveness to entire polypeptides can be influenced by the thymus-directed process of shaping an individual's T cell repertoire. As such, it seems doubtful that the strategy of employing solitary cloned proteins of large viruses such as HSV for vaccination will prove adequate.

Vaccination schemes involving individual polypeptides are further complicated by the fact that T lymphocytes are potentially capable of recognizing any virally encoded gene product. Evidence that internal antigens could be recognized by CTLs first began to surface in studies of the influenza system (BRACIALE 1987; BIDDISON et al. 1987), in which a component of the CTL response was defined as being targeted against matrix proteins. More recently, internal proteins of some viruses have been shown to be major target antigens for CTLs (JONJIC et al. 1988; TOWNSEND et al. 1985; MARTIN et al. 1990; BANKS et al. 1991). For HSV, murine CTLs specific for ICP4 have been demonstrated, but CTLs (at least from H-2^k mice) fail to recognize ICPO (MARTIN et al. 1990). One report describing the in vitro response to cytomegalovirus showed that the majority of the CTL response was directed against an immediate early protein of that virus (JONJIC et al. 1988). A study of the HSV-1 immediate early protein ICP27 revealed that is could be recognized by herpes-specific CTLs generated in H-2^d mice but not H-2^k or H-2^b mice (BANKS et al. 1991). It was also shown that some elements of specific immunity, including CTLs, could be induced by ICP27 but that the immune response induced by this protein could not protect against all forms of HSV-1 challenge in vivo.

On the basis of these observations, it seems clear that any vaccination strategy which has the goal of eliciting specific CTL activity must provide an exposure to as many HSV proteins as feasible.

3.1.2 In Vivo Studies

HSV-specific CTLs have been demonstrated to substantially curtail the progression of HSV infections in vivo by adoptively transferring defined populations of cells into naive syngeneic animals. The technically simplest, and perhaps the least clinically relevant approach, to conducting these kind of studies has examined the ability of adoptively transferred killer cells to protect recipient mice from a lethal challenge dose of infectious virus (LARSEN et al. 1983). The current trend is toward adoptive transfer models which better reflect human HSV infection.

The two most commonly employed systems are the ear pinna and footpad models, in which either the ability of a defined population of cells to clear a local inoculum of virus is measured (LEUNG et al. 1984; MARTIN and ROUSE; ALLEN et al. 1991, 1987), or reductions in the establishment of latent infections (BONNEAU and JENNINGS 1989) are assessed by the ability to recover virus from explanted dorsal root ganglia.

A controversy arose early in the course of HSV T cell adoptive transfer studies over which subpopulation of T cells was chiefly responsible for mediating the observed in vivo antiviral effects. At least four different groups studied the effect of adoptively transferring in vitro cultured, HSV-specific T cell populations into HSV-challenged mice (SETHI et al. 1983; LARSEN et al. 1983; NASH and GELL 1983; SCHRIER et al. 1983). Of these several independently working laboratories, two concluded that the $CD4^+$ subpopulation of T cells was conferring protection in vivo, with little or no contribution from $CD8^+$ T cells (NASH and GELL 1983; SCHRIER et al. 1983). The other investigators came to precisely the opposite conclusion, namely, that $CD8^+$ T cells were exclusively responsible for protecting mice in adoptive transfer experiments (SETHI et al. 1983; LARSEN et al. 1983). Further experimentation revealed that both $CD4^+$ and $CD8^+$ T cell populations, which had been expanded in culture with a secondary exposure to antigen, were capable of protecting recipient mice from HSV challenges (MARTIN and ROUSE 1987; NASH et al. 1987). However, while these more recent studies verify that protection can be mediated against HSV infection by either major population of T cells, they fail to indicate why the earlier discrepancies occurred. The answer to that question may lie in how the transferred cells were exposed to antigen, both for the primary in vivo inoculation and for the secondary in vitro stimulation. It has been known for some time that a condition of split tolerance can be induced in mice, in which normal CTL, antibody, and helper cell responses are stimulated against HSV but with a dramatic impairment of T cells capable of mediating delayed-type hypersensitivity reactions (NASH and ASHFORD 1982; LATHEY et al. 1987). The induction of split tolerance in these animals appears to be dependent upon the route of antigen injection (NASH and ASHFORD 1982). It is now clear that different modes of antigen processing preferentially activate different subpopulations of T cells, a phenomenon which has been observed in the human HSV CTL response (YASUKAWA et al. 1989). The exogenous pathway, in which soluble antigen is endocytosed and processed by phagocytic cells, tends to activate class II-restricted T lymphocytes. In contrast, the endogenous pathway, in which the viral proteins produced during an active infection are processed and presented on the infected cell itself, tends to activate class I-restricted T cells. Thus, it may be that the observed differences in the early adoptive transfer studies were attributable entirely to the means by which the injected lymphocyte populations were initially activated and expanded in culture. This notion is supported by experiments (LARSEN et al. 1984) which demonstrated that the subpopulations involved in controlling HSV infection varied in accordance with the stage and the nature of the virus:host interaction. Recently, NASH and coworkers (1987) showed that a cloned $CD4^+$ T cell line was very effective at resolving a localized HSV infection of the ear pinna when injected intravenously into mice. The same study also presented evidence that these cells could reduce the incidence at which latent infection was established in test animals. As such, although the split tolerance experiments demonstrated that class II-restricted T cells are not essential to the successful containment of HSV infections, they are nevertheless effective when they are available.

Another aspect of the cell activation status of transferred populations that needs to be considered is the level of expression of homing-associated adhesion molecules. Several studies have demonstrated the preferential accumulation of lymphocyte subsets, expressing high levels of various integrin family molecules, to inflammatory sites (BERG et al. 1989; PITZALIS et al. 1988; PICKER et al. 1990). Similar differential infiltration patterns have been observed directly in human lesions (CUNNINGHAM et al. 1985), which are characterized by an early accumulation of $CD4^+$ T cells followed later in time by $CD8^+$ T cells. As such, studies using adoptive transfer need to take into account the possibility that in vitro expansion protocols may result in populations which, while efficacious at clearing infections in recipients, may have little or no in vivo relevance in the control of natural infections.

Balb/c and CBA mice inoculated with gD-transfected cells (BLACKLAWS and NASH 1990) or with a gD-expressing vaccinia virus construct (MARTIN and ROUSE 1987) mount a protective response which leads to rapid clearance of local challenges and decreases the incidence of latent infection; gB also has this effect, although to a lesser extent (BLACKLAWS and NASH 1990). The response to gD has, in most systems, been limited to antibody and delayed-type hypersensitivity phenotype T cells, leading some to believe that a CTL response is not required for the resolution of HSV lesions. In fairness, however, the cytolytic potential of these cells has almost never been studied in the murine system, in which investigators have long made extensive use of Ia^- target cells. It is only recently that murine studies have begun to assess HSV-specific $CD4^+$ T cells for their ability to lyse target cells. At least three studies to date have reported HSV-specific CTL activity in the mouse, which is mediated by $CD4^+$ (JOHNSON et al. 1990; MARTIN et al. 1987; KOLAITIS et al. 1990). In one of those reports (KOLAITIS et al. 1990), approximately 30% of the observed CTL activity was attributed to $CD4^+$ T cells. Considering that class II-restricted CTLs are commonly observed in peripheral blood from HSV-seropositive humans (YASUKAWA and ZARLING 1984a; SCHMID 1988; TORPEY et al. 1989) and that immunopathologic studies have implicated these same cells as having a central role in the clearance of local HSV lesions (CUNNINGHAM et al. 1985), it may be that $CD4^+$ CTLs are providing a similar function in the murine model.

3.2 Guinea Pig Model

Female guinea pigs infected vaginally with HSV develop a limited infection, establish latency in the local sensory ganglia, and subsequently develop recurrent disease (STANBERRY et al. 1985). As such, the course of genital HSV infection in guinea pigs precisely parallels the disease state as it occurs in humans. This has made it an extremely useful model for assessing the efficacy of vaccine candidates (STANBERRY et al. 1988, 1989; HO et al. 1989; BERNSTEIN et al. 1988) and antiviral agents (STANBERRY et al. 1986, 1990) in the control of HSV infections. The study of humoral immunity to HSV has been fairly well characterized in the guinea pig (BERNSTEIN et al. 1988; HO et al. 1989; STANBERRY et al. 1989), but definitive examinations of the T cell-mediated immune response have not been possible until recently (MCBRIDE et al. 1989). There are now available both monoclonal antibodies which identify T lymphocytes in the guinea pig and a CTL marker (TAN et al. 1985). The advent of these critical reagents should propel CTL studies in this model system in the coming years. One report has already demonstrated a potent T cell response to the Skinner vaccine, which is a formalin-extracted HSV subunit vaccine (MBBRIDE et al. 1989). The chief limitations of this model remain the unavailability of inbred syngeneic strains, which are essential to adoptive transfer studies, and the lack of information on the MHC of guinea pigs.

4 In Vivo Role of $CD4^+$ CTLs

Since 1982, the CTL responses to a number of viruses and to some bacterial pathogens have been shown to contain a large component of activity which is restricted to class II MHC antigens and mediated by $CD4^+$ T cells. This list includes, but is not limited to, HSV types 1 and 2 (YASUKAWA and ZARLING 1984a, b; SCHMID 1988), varicella zoster virus (HAYWARD et al. 1986; ARVIN 1991), influenza (FLEISCHER et al. 1985; KAPLAN et al. 1984), cytomegalovirus (LINDSLEY et al. 1986), measles (JACOBSON et al. 1984, 1989), and human pathogenic species of the genus Mycobacterium (BARNES et al. 1989; OTTENHOFF and MUTIS 1990). Despite general agreement that $CD4^+$ CTL activity is readily induced and that these cells are present at high frequency in the peripheral blood of seropositive donors, a direct combat role for these cells in the elimination of infectious agents is still widely unaccepted. The principal objection to their participation is that very few cell types within the body normally express class II MHC antigens and, as such, could not be recognized as target cells by $CD4^+$ CTLs. This argument has led some investigators to propose a limited regulatory role for $CD4^+$ CTLs, in which these cells destroy antigen-presenting cells and class II antigen-positive CTLs, resulting in the down-regulation of an ongoing response (BRAAKMAN et al. 1987; OTTENHOFF and MUTIS 1990). It is probable that $CD4^+$ CTLs perform this type of regulatory role, but this by no means precludes a role for these cells in the elimination or control of some disease states.

There are some differences between $CD4^+$ and $CD8^+$ CTLs, for example the requirements for induction (RUDD et al. 1989) and for antigen presentation

(BRACIALE et al. 1987). It has been suggested by some investigators that CD4$^+$ CTLs use only soluble factors to kill targets and that CD8$^+$ CTLs kill using other means, i.e. those more directly dependent on cognitive processes between effector cell and target (TITE and JANEWAY 1984). However, it has been shown that CD4$^+$ CTLs kill with kinetics that are indistinguishable from those of CD8$^+$ cells (SCHMID et al. 1986). Furthermore, CD4$^+$ killers produce cytolytic granules in response to antigen activation, exactly like CD8$^+$ killer cells (RUDDLE et al. 1987). Finally, it was demonstrated that CD8$^+$ CTLs can kill bystander targets (RUDDLE and HOMER 1987) and produce the same array of secreted cytokines under antigen stimulation as the Th1 subset of CD4$^+$ T cells (FONG and MOSMANN 1990). Thus, while certain properties of these two subpopulations are inevitably dissimilar, it seems equally certain that both cell types utilize the same fundamental mechanisms for effecting target cell lysis. Furthermore, the unique properties that separate the CD4$^+$ and CD8$^+$ populations, which chiefly involve the conditions under which these subsets are activated, probably help define the disease states for which CD4$^+$ vs CD8$^+$ CTLs will predominate.

As to the apparent difficulty in finding suitable target cells, it has been known for some time that IFN-γ is capable of inducing the expression of high levels of MHC class II antigens on a variety of cell types (HALLORAN et al. 1986; SOLLID et al. 1987; GROENEWEGEN et al. 1987; KALLENBERG et al. 1987; CRESSWELL 1987). In addition, IFN-γ is produced in response to virtually all viral infections, including HSV (GLASGOW 1974; RASMUSSEN et al. 1974).

Immunopathologic analysis of HSV lesions sampled in serial biopsy revealed that class II antigen expression on keratinocytes begins within 24 h (CUNNINGHAM et al. 1985). This observation has recently been verified by immunohistologic studies of corneal buttons excised from herpes keratitis patients (EL-ASRAR et al. 1990). By the end of 48 h, class II antigen expression is uniform on all keratinocytes within and adjacent to the lesion. Moreover, this same study revealed that only CD4$^+$ CTLs can be found as infiltrating lymphocytes during the first 48 h following prodrome. After that point, CD8$^+$ T cells also begin to infiltrate and eventually are found in the same proportion with respect to CD4$^+$ T cells as is found in peripheral blood. NK cells are underrepresented in HSV lesions throughout the course of lytic infection.

Typical HSV lesions have already begun to resolve in 48 h and little or no infectious virus can be isolated from them (CUNNINGHAM et al. 1985). As such, it seems reasonable to suppose a central role for CD4$^+$ T cells in the control of recurrent lesions. A more recent study revealed that HSV-infected autologous keratinocytes, converted in vitro by IFN-γ to express class II antigens, are susceptible to killing by CD4$^+$ CTLs (CUNNINGHAM and NOBLE 1989). These class II-restricted cells are likelly to provide a variety of functions, ranging from the activation and attraction of phagocytic cells to direct cytolytic action to the attraction of CD8$^+$ CTLs. The late infiltrating CD8$^+$ cells, which can also be identified at high frequency in peripheral blood when antigen is presented by the endogenous pathway (YASUKAWA et al. 1989), probably assist in the final stages of resolution.

These observations in human lesions are supported by the adoptive transfer studies in mice, in which various groups have demonstrated both $CD4^+$ and $CD8^+$ T cells to be effective in curtailing herpesvirus infections.

It seems reasonable to speculate that the disease environment most conducive to the generation of cytolytic $CD4^+$ CTLs is probably the local cutaneous lesion, which would account for the observation of high numbers of antigen-specific $CD4^+$ CTLs in individuals who have been infected with HSV (SCHMID 1988), varicella zoster virus (ARVIN et al. 1991), and measles (SCHMID, unpublished observation).

5 Conclusions

A considerable body evidence exists which favors an important role for T cell-mediated immunity in the control of both primary and recurrent HSV infections. While much of the evidence from human studies is, for ethical reasons, circumstantial, the wealth of data now available from clinical studies, in vitro assay systems, and in particular immunohistologic analyses of HSV lesions argue persuasively for a central role of CTLs in the resolution of HSV lesions. The continued refinement of animal models into ever closer approximations of human disease adds considerable weight to that argument. As such, treatment modalities which cause the activation of CTLs need to be given due consideration in the design of HSV vaccines.

Acknowledgment. This work was supported by Public Health Services Grants EY05093 and AI14981.

References

Allen EM, Weir JP, Martin S, Mercadal C, Rouse BT (1990) Role of coexpression of IL-2 and herpes simplex virus proteins in recombinant vaccinia virus vectors on levels of induced immunity. Viral Immunol 3: 207–215

Arvin AA, Sharp M, Smith S, Koropchak CM, Diaz PS, Kinchington P, Ruyechan W, Hay J (1991) Equivalent recognition of a varicella-zoster virus immediate early protein (IE62) and glycoprotein I by cytotoxic T lymphocytes of either $CD4^+$ or $CD8^4$ phenotype. J Immunol 146: 257–264

Banks TA, Allen EM, Dasgupta S, Sandri-Goldin R, Rouse BT (1991) Herpes simplex virus type I-specific cytotoxic T lymphocytes recognize immediate early protein ICP27. J Virol 65: 3185–3191

Barnes PF, Mistry SD, Cooper CL, Pirmez C, Rea TH, Modlin RL (1989) Compartmentalization of a $CD4^+$ T lymphocyte subpopulation in tuberculosis pleuritis. J Immunol 142: 1114–1119

Berg EL, Goldstein LA, Jutila MA, Nakache M, Picker LJ, Streeter PR, Wu, NW, Zhou D, Butcher EC (1989) Homing receptors and vascular addressins: cell surface molecules that direct lymphocyte traffic. Immunol Rev 108: 5–18

Bernstein DI, Stanberry LR, Kappes JC, Burke RL, Myer MG (1988) Antibody to herpes simplex virus (HSV) ICP-35 proteins after HSV challenge of animals immunized with HSV subunit vaccine. J Infect Dis 157: 1178–1186

Bertagnolli M, Herrmann S (1990) IL-7 supports the generation of cytotoxic T lymphocytes from thymocytes. Multiple lymphokines required for proliferation and cytotoxicity. J Immunol 145: 1706–1712
Biddison WE, Doherty PC, Webster RG (1977) Antibody to influenza virus matrix protein detects a common antigen on the surface of cells infected with type A influenza viruses. J Exp Med 146: 690–697
Blacklaws BA, Nash AA (1990) Immunological memory to herpes simplex virus type 1 glycoproteins B and D in mice. J Gen Virol 71: 863–871
Blacklaws BA, Nash AA, Darby G (1987) Specificity of the immune response ot mice to herpes simplex virus glycoproteins B and D constitutively expressed on L cell lines. J Gen Virol 68: 1103–1114
Bodmer JM, Bastin JM, Askonas BA, Townsend ARM (1989) Influenza-specific cytotoxic T-cell recognition is inhibited by peptides unrelated in both sequence and MHC restriction. Immunology 66: 163–169
Bonneau RH, Jennings RH (1989) Modulation of acute and latent herpes simplex virus infection in C57BL/6 mice by adoptive transfer of immune lymphocytes with cytolytic activity. J Virol 63: 1480–1484
Braakman E, Rotteveel FTM, van Bleek G, van Seventer GA, Lucas CJ (1987) Are MHC class II-restricted cytotoxic T lymphocytes important? Immunol Today 8: 265–267
Braciale TJ (1977) Immunologic recognition of influenza virus-infected cells. II. Expression of influenza A matrix protein on the infected cell surface and its role in recognition by cross-reactive cytotoxic T cells. J Exp Med 146: 673–690
Braciale TJ, Morrison LA, Sweetser MT, Sambrook J, Gething M-J, Braciale VL (1987) Antigen presentation pathways to class I and class II MHC-restricted T lymphocytes. Immunol Rev 98: 95–114
Corey L, Spear PG (1986) Infections with herpes simplex viruses. N Engl J Med 314: 686–691
Cresswell P (1987) Regulation of HLA class I and class II antigen expression. Br Med Bull 43: 66–80
Cunningham AL, Noble JR (1989) Role of keratinocytes in human recurrent herpetic lesions. Ability to present herpes simplex virus antigen and act as target for T lymphocyte cytotoxicity in vitro. J Clin Invest 83: 490–496
Cunningham AL, Turner RR, Miller AC, Para MF, Merigan TC (1985) Evolution of recurrent herpes simplex lesions: an immunolhistologic study. J Clin Invest 75: 226–233
Douglas RG, Couch RB (1970) A prospective study of chronic herpes virus infection and recurrent herpes labialis in humans. J Immunol 104: 289–295
Eberle R, Russell RG, Rouse BT (1981) Cell mediated immunity to herpes simplex virus: recognition of type specific and type common surface antigens by cytotoxic T cell populations. Infect Immun 34: 795–803
El-Asrar AM, Geboes K, Missotten L, Emarah MH, Desmet V (1990) Expression of MHC class II antigens and immunoglobulin M by the corneal epithelial cells in herpetic keratitis. Int Ophthalmol 14: 233–239
Farrar JJ, Simon PL, Farrar WL, Keopman J, Fuller-Bonar J (1981) Role of mitogenic factor, lymphocyte activating factor, and immune interferon in the induction of humoral and cell-mediated immunity. Ann NY Acad Sci 332: 303–315
Farrar WL, Johnson HM, Farrar JJ (1982) Regulation of the production of immune interferon and cytotoxic T lymphocytes by interleukin 2. J Immunol 126: 1120–1125
Fleischer B, Becht H, Rott R (1985) Recognition of viral antigens by human influenza A virus-specific T lymphocyte clones. J Immunol 135: 2800–2804
Fong TAT, Mosmann TR (1990) Alloreactive muring $CD8^+$ T cell clones secrete the Th1 pattern of cytokines. J Immunol 144: 1744–1752
Glasgow LA (1974) Cytomegalovirus interference in vitro. Infect Immun 9: 702–707
Glorioso J, Kees U, Kümel G, Kirchner H, Krammer PH (1985) Identification of herpes simplex virus type 1 (HSV-1) glycoprotein gC as the immunodominant antigen for HSV-1-specific memory cytotoxic T lymphocytes. J Immunol 135: 575–582
Groenewegen G, Buurman WA, de Ley M, Kootstra G (1987) A new lymphokine different from interferon-γ induces major histocompatibility complex class II antigen expression. Transplant Proc 19: 190–191
Halloran PF, Wadgymar A, Autenreid P (1986) The regulation of expression of major histocompatibility complex products. Transplantation 41: 413–420
Hayward AR, Pontesilli O, Herberger M, Laszlo M, Levin M (1986) Specific lysis of varicella zoster virus-infected B lymphoblasts by human T cells. J Virol 58: 179–184

Ho M (1977) Viral infections after transplantation in man. Arch Virol 55: 1–24
Ho RJY, Burke RL, Merigan TC (1989) Antiven-presenting liposomes are effective in treatment of recurrent herpes simplex virus genitalis in guinea pigs. J Virol 63: 2951–2959
Horohov DW, Wyckoff JH, Moore RN, Rouse BT (1986) Regulation of herpes simplex virus-specific cell-mediated immunity by a specific suppressor factor. J Virol 58: 331–338
Jacobson S, Richert JR, Biddison WE, Satinsky A, Hartzmann RJ, McFarland HF (1984) Measles virus specific T4^{+} human cytotoxic T-cell clones are restricted by class II HLA antigens. J Immunol 133: 754–757
Jacobson S, Sekaly RP, Jacobson CL, McFarland HF, Long EO (1989) HLA class II-restricted presentation of cytoplasmic measles virus antigens to cytotoxic T cells. J Virol 63: 1756–1762
Johnson RM, Lancki DW, Fitch FW, Spear PG (1990) Herpes simplex virus glycoprotein D is recognized as antigen by CD4^{+} and CD8^{+} T lymphocytes from infected mice. Characterization of T cell clones. J Immunol 145: 702–710
Jonjic S, del Val M, Keil GM, Reddehase MJ, Koszinowski UH (1988) A nonstructural viral protein expressed by recombinant vaccinia virus protects against lethal cytomegalovirus infection. J Virol 62: 1653–1658
Kallenberg CGM, Schillizi BM, Beaumont F, Poppema S, De Leij L, The TH (1987) Expression of class II MHC antigens on alveolar epithelium in fibrosing alveolitis. Clin Exp Immunol 67: 182–190
Kaplan DR, Griffith R, Braciale VL, Braciale TJ (1984) Influenza virus-specific human cytotoxic T cell clones: heterogeneity in antigenic specificity and restriction by class II MHC products. Cell Immunol 88: 193–206
Kieff E, Hoyer B, Bachenheimer S, Roizman B (1972) Genetic relatedness of type 1 and type 2 herpes simplex viruses. J Virol 9: 738–745
Kohl S (1984) The immune response of the neonate to herpes simplex virus infection. In: Rouse BT, Lopez C (eds) Immunobiology of herpes simplex virus infection. CRC Press. Boca Raton
Kolaitis G, Doymaz M, Rouse BT (1990) Demonstration of MHC class II-restricted cytotoxic T lymphocyts in mice against herpes simplex virus. Immunology 71: 101–106
Larsen HS, Russel RG, Rouse BT (1983) Recovery from lethal herpes simplex virus type 1 infection is mediated by cytotoxic T lymphocytes. Infect Immun 41: 197–204
Larsen HS, Feng M-F, Horohov DW, Moore RN, Rouse BT (1984) Role of T-lymphocyte subsets in recovery from herpes simplex virus infection. J Virol 50: 56–59
Lathey JL, Martin S, Rouse BT (1987) Suppression of delayed type hypersensitivity to herpes simplex virus type 1 following immunization with an anti-idiotypic antibody: an example of split tolerance. J Gen Virol 68: 1093–1102
Lawman MJP, Courtney RJ, Eberle R, Schaffer PA, O'hara MK, Rouse BT (1980a) Cell-mediated immunity to herpes simplex virus: specificity of cytotoxic T cells. Infect Immun 30: 451–461
Lawman MJP, Rouse BT, Courtney RJ, Walker RD (1980b) Cell-mediated immunity against herpes simplex induction of cytotoxic T lymphocytes. Infect Immun 27: 133–139
Leung KN, Nash AA, Sia DY, Wildy P (1984) Clonal analysis of T cell responses to herpes simplex virus: isolation, characterization and antiviral properties of an antigen-specific helper T-cell clone. Immunology 53: 623–633
Lindsley MD, Torpey DJ III, Rinaldo CR Jr (1986) HLA-DR-restricted cytotoxicity of cytomegalovirus-infected monocytes mediated by leu-3-positive T cells. J Immunol 136: 3045–3051
Lopez C (1984) Natural resistance mechanisms against herpesvirus in health and disease. In: Immunobiology of herpes simplex virus infection. CRC Press, Boca Raton
Martin S, Rouse BT (1987) The mechanism of antiviral immunity induced by a vaccinia recombinant expressing herpes simplex virus type 1 glycoprotein D. II. Clearance of local infection. J Immunol 138: 3431–3437
Martin S, Moss B, Berman PW, Laskey LA, Rouse BT (1987) Mechanisms of antiviral immunity induced by a vaccinia virus recombinant expressing herpes simplex virus type 1 glycoprotein D: cytotoxic T cells. J Virol 61: 726–734
Martin S, Xiuxuan Z, Silverstein SJ, Courtney RJ, Yao F, Jenkins FJ, Rouse BT (1990) Murine cytotoxic T lymphocytes specific for herpes simplex virus type 1 recognize the immediate early protein ICP4 but not ICPO. J Gen Virol 71: 2391–2399
McBride BW, Ridgeway P, Phillpotts R, Newell DG (1989) Systemic and mucosal cell mediated immune responses against guinea-pigs detected by lymphocyte proliferation and monoclonal antibodies against macrophages and T cells. Vaccine 7: 409–416
Merigan TC, Stevens DA (1971) Viral infections in man associated with acquired immunological deficiency states. Fed Proc 30: 1858–1864

Mester JC, Highlander SL, Osmand AP, Glorioso JC, Rouse BT (1990) Herpes simplex virus type 1-specific immunity induced by peptides corresponding to an antigenic site of glycoprotein B. J Virol 64: 5277–5283
Meuer SC, Acuto O, Hercend T, Schlossman SF, Reinherz EL (1984) The human T cell receptor. Ann Rev Immunol 2: 23–50
Marahan PS (1984) Interactions of herpesviruses with mononuclear phagocytes. In: Immunobiology of herpes simplex virus infection. CRC Press, Boca Raton
Nash AA, Ashford NPN (1982) Split T-cell tolerance in herpes simplex virus-infected mice and its implication for anti-viral immunity. Immunology 45: 761–767
Nash AA, Gell PGH (1983) Membrane phenotype of murine effector and suppressor T cells involved in delayed hypersensitivity and protective immunity to herpes simplex virus. Cell Immunol 75: 348–355
Nash AA, Jayasuriya A, Phelan J, Cobbold SP, Waldmann H, Prospero T (1987) Different roles for $L3T4^+$ and $Lyt2^+$ T cell subsets in the control of an acute herpes simplex virus infection of the skin and nervous system. J Gen Virol 68: 825–833
Ottenhoff THM, Mutis T (1990) Specific killing of cytotoxic T cells and antigen-presenting cells by $CD4^+$ cytotoxic T cell clones. A novel potentially immunoregulatory T–T cell interaction in man. J Exp Med 171: 2011–2024
Pass RF, Whitley RJ, Whelchel JD, Diethelm AG, Reynolds DW, Alford CA (1979) Identification of patients with increased risk of infection with herpes simplex virus after renal transplantation. J Infect Dis 140: 487–492
Pfizenmaier K, Starzinski-Powitz A, Röllinghoff M, Falke D, Wagner H (1977) T cell mediated cytotoxicity against herpes simplex virus-infected target cells. Nature 265: 630–632
Picker LJ, Terstappen LWMM, Rott LS, Streeter PR, Stein H, Butcher EC (1990) Differential expression of homing-associated adhesion molecules by T cell subsets in man. J Immunol 145: 3247–3255
Pitzalis C, Kingsley G, Haskard D, Panayi G (1988) The preferential accumulation of helper-inducer T lymphocytes in inflammatory lesions: evidence for regulation by selective endothelial and homotypic adhesion. Eur J Immunol 18: 1397–1404
Prymowicz D, Moore RN, Rouse BT (1985) Frequency of herpes simplex virus-specific helper T lymphocyte precursors in the lymph node cells of infected mice. J Immunol 134: 2683–2688
Quinnan GV, Masur H, Rook AH, Armstrong G, Frederick WR, Epstein J, Manischewitz MS, Mecher AM, Jackson L, Ames J, Smith HA, Parker M, Pearson GR, Panillo J, Mitchell C, Strauss SE (1984) Herpesvirus infections in the acquired immune deficiency syndrome. JAMA 252: 72–77
Rand KH, Rasmussen LE, Pollard RB, Arvin AA, Merigan TC (1976) Cellular immunity and herpesvirus infections in cardiac transplant patients. N Engl J Med 296: 1372–1377
Rasmussen LE, Jordan GW, Stevens DA, Merigan TC (1974) Lymphocyte interferon production and transformation after herpes simplex infections in humans. J Immunol 112: 728–736
Raulet D, Bevan MJ (1982) A differentiation factor required for the expression of cytotoxic T cell function. Nature 296: 754–757
Rosenthal KL, Smiley JR, South S, Johnson DC (1987) Cells expressing herpes simplex virus glycoprotein gC but not gB, gD, or gE are recognized by murine virus-specific cytotoxic T lymphocytes. J Virol 61: 2438–2447
Rouse BT (1984a) Cell-mediated immune mechanisms. In: Immunobiology of herpes simplex virus infection. CRC Press, Boca Raton
Rouse BT (1984b) Role of adaptive immune defense mechanisms in herpes simplex resistance. In: Immunobiology of herpes simplex virus infection. CRC Press, Boca Raton
Rouse BT, Lawman MJP (1980) Induction of cytotoxic T lymphocytes against herpes simplex virus type 1: role of accessory cells and amplifying factor. J Immunol 124: 2341–2346
Rouse BT, Wagner H (1984) Frequency of herpes simplex virus-specific cytotoxic T lymphocyte precursors in lymph node cells of infected mice. Immunology 51: 57–64
Rudd CE, Anderson P, Morimoto C, Streuli M, Schlossman SF (1989) Molecular interactions, T-cell subsets and a role of the CD4/CD8: $p56^{lck}$ complex in human T-cell activation. Immunol Rev 111: 226–265
Ruddle NH, Homer R (1987) The role of lymphotoxin in inflammation. Prog Allergy 40: 162–182
Ruddle NH, McGrath K, James T, Schmid DS (1987) Purified lymphotoxin (LT) from class I restricted CTLs and class II restricted cytolytic helpers induce target cell DNA fragmentation. In: Bonavida B, Collier RJ (eds) Membrane-mediated cytotoxicity. Liss, New York
Schmid DS (1988) The human MHC-restricted cellular response to herpes simplex virus type 1 is mediated by $CD4^+$, $CD8^-$ T cells and is restricted to the DR region of the MHC complex. J Immunol 140: 3610–3616

Schmid DS, Rouse BT (1983) Cellular interactions in the cytotoxic T lymphocyte response to herpes simplex virus antigens: differential antigen activation requirements for the helper T lymphocyte and cytotoxic T lymphocyte precursors. J Immunol 131: 479–484

Schmid DS, Larsen HS, Rouse BT (1981) The role of accessory cells and T cell-growth factor in induction of cytotoxic T lymphocytes against herpes simplex virus antigens. Immunology 44: 735–763

Schmid DS, Larsen HS, Rouse BT (1982) Role of Ia antigen expression and secretory function of accessory cells in the induction of cytotoxic T lymphocyte responses against herpes simplex virus. Infect Immun 37: 1138–1147

Schmid DS, Tite JP, Ruddle NH (1986) DNA fragmentation: manifestation of target cell destruction mediated by cytotoxic T-cell lines, lymphotoxin-secreting helper T cell clones, and cell-free lymphotoxin-containing supernatant. Proc Natl Acad Sci USA 83: 1881–1885

Schrier RD, Pizer LI, Moorhead JW (1983) Type-specific delayed hypersensitivity and protective immunity induced by isolated herpes simplex glycoprotein. J Immunol 130: 1413–1418

Sethi KK, Brandis H (1977) Specifically immune mouse T cells can destroy H-2 compatible murine target cells infected with herpes simplex virus types 1 and 2. Z Immunitätsforsch 150: 162–173

Sethi KK, Omata Y, Schneweis KE (1983) Protection of mice from fatal herpes simplex virus type 1 infection by adoptive transfer of cloned virus-specific and H-2 restricted cytolytic T lymphocytes. J Gen Virol 64: 443–447

Shore SL, Nahmias A (1981) Immunobiology of herpes simplex viruses. In: Nahmias AJ, O'Reill (eds) Immunology of human infections. Plenum, New York

Siegal FP, Lopez C, Hammer GS, Brown AE, Kornfeld SJ, Gold J, Hassett J, Hirshman SZ, Cunningham-Rundles C, Adelsberg BR, Parham DM, Siegal M, Cunningham-Rundles S, Armstrong D (1981) Severe acquired immunodeficiency in male homosexuals manifested by chronic herpes simplex lesions. N Engl J Med 305: 1439–1444

Sollid LM, Gaudernack G, Markussen G, Kvale D, Brantzaeg P, Thorsby E (1987) Induction of various HLA class II molecules in a human colonic adenocarcinoma cell line. Scand J Immunol 25: 175–180

Spencer ES, Anderson HK (1970) Clinically evident nonterminal infections with herpes simplex viruses and the wart virus in immunosuppressed renal allograft patients. Br Med J iii: 251–254

Stanberry LR, Kern ER, Richards JT, Overall JC Jr (1985) Recurrent genital herpes simplex virus infection in guinea pigs. Intervirology 24: 226–231

Stanberry LR, Bernstein DI, Myers MG (1986) Evaluation of the herpes simplex virus antiviral activity of pyrethrins. Antiviral Res 6: 95–102

Stanberry LR, Burke RL, Myers MG (1988) Herpes simplex virus glycoprotein treatment of recurrent genital herpes. J Infect Dis 157: 156–163

Stanberry LR, Myers MG, Stephanopoulos DE, Burke RL (1989) Preinfection prophylaxis with herpes simplex virus glycoprotein immunogens: factors influencing efficacy. J Gen Virol 70: 3177–3185

Stanberry LR, Harrison CJ, Bravo FJ, Childs F, Reece AL, Bernstein DI (1990) Recurrent genital herpes in the guinea pig augmented by ultraviolet radiation: effects of treatment with acyclovir. Antiviral Res 13: 227–236

Tan BTG, Ekelaar F, Lourink J, Rimmelzwaan G, DeJonge AJR, Scheper RJ (1986) Production of monoclonal antibodies defining guinea pig T cell surface markers and a strain 13 Ia-like antigen: the value of immunohistological screening. Hybridoma 4: 115–124

Taylor PM, Davey J, Howland K, Rothbard J, Askonas BA (1987) Class I MHC molecules, rather than other mouse genes dictate influenza epitope recognition by cytotoxic T cells. Immunogenetics 26: 267–271

Tite JP, Janeway CA Jr (1984) Cloned helper T cells can kill B lymphoma cells in the presence of specific antigen: Ia restriction and cognate vs. noncognate interactions in cytolysis. Eur J Immunol 14: 878–886

Torpey DJ III, Lindsley MD, Rinaldo CR Jr (1989) HLA-restricted lysis of herpes simplex virus-infected monocytes and macrophages mediated by $CD4^+$ and $CD8^+$ T lymphocytes. J Immunol 142: 1325–1332

Townsend ARM, Gotch FM, DAvey J (1985) Cytotoxic T cells recognize fragments of nucleoprotein. Cell 42: 457–467

Townsend ARM, Rothbard J, Gotch FM, Bahadar G, Wraith D, McMichael AJ (1986) The epitopes of influenza nucleoprotein recognized by cytotoxic T lymphocytes can be defined with short synthetic peptides. Cell 44: 959–968 van Binnendijk RS, Poelen MCM, De Vries P, Voorma HO, Osterhaus ADME, Uytdehaag KGCM (1989) Measles virus-specific human T cell clones. Characterization of specificity and function of $CD4^+$ helper/cytotoxic and $CD8^+$ cytotoxic T cell clones. J Immunol 142: 2847–2854

Wagner H, Hardt C, Rouse BT, Röllinghoff M, Scheurich P, Pfizenmaier K (1982) Dissection of the proliferative and differentiating signals controlling murine cytotoxic T lymphocyte response. J Exp Med 155: 1876–1881

Whitley RJ (1985) Epidemiology of herpes simplex viruses. In: Roizman B (ed) herpesviruses, vol 3. Plenum, New York

Witmer LA, Rosenthal KL, Graham FL, Friedman HM, Yee A, Johnson DC (1990) Cytotoxic T lymphocytes specific for herpes simplex virus (HSV) studied using adenovirus vectors expressing HSV glycoproteins. J Gen Virol 71: 387–396

Yasukawa M, Zarling JM (1984a) Human cytotoxic T cell clones directed against herpes simplex virus-infected cells. I. Lysis restricted by HLA class II MB and DR antigens. J Immunol 133: 422–427

Yasukawa M, Zarling JM (1984b) Human cytotoxic T cell clones directed against herpes simplex virus-infected cells. II. Bifunctional clones with cytotoxic and virus-induced proliferative activities exhibit herpes simplex virus type 1 and type 2 specific or type common reactivities. J Immunol 133: 2736–2742

Yasukawa M, Inatsuki A, Kobayashi Y (1989) Differential in vitro activation of $CD4^+CD8^-$ and $CD8^+CD4^+$ herpes simplex virus-specific human cytotoxic T cells. J Immunol 143: 2051–2057

Zinkernagel R, Doherty PC (1979) MHC-restricted cytotoxic T cells: studies on the biological role of polymorphic major transplantation antigens determining T-cell restriction-specificity, function, and responsiveness. Adv Immunol 27: 51–177

Zweerink HJ, Stanton LW (1981) Immune response to HSV infections: virus specific antibodies in sera from patients with recurrent facial infections. Infect Immun 31: 624–630

Zweerink HJ, Askonas BA, Millican D, Courtneidge SA, Skehel JJ (1977) Cytotoxic T cells to type A influenza virus: viral hemagglutinin induces A-strain specificity while infected cells confer cross-reactive cytotoxicity. Eur J Immunol 7: 630–635

The Role of Antibody in Herpes Simplex Virus Infection in Humans

S. KOHL

1 Introduction

This chapter will attempt to critically analyze the role of antibody in the host's response to herpes simplex virus (HSV) infection. Although the emphasis will be on the human, where relevant, selected animal data will be included.

There are a large variety of measured antibodies to HSV. Older work tends to center around antibodies with functions such as complement fixation, hemagglutination inhibition, and viral neutralization and fluorescently labeled antibodies of various types. More recent work has utilized whole fixed virus or virus components to determine the presence of antibody by using solid phase assays such as enzyme-linked immune adsorption (ELISA), radioimmunoassay, or western blot techniques. These assays have the advantage of being able to determine the antibody response to specific portions of the virus, such as the various glycoproteins found in the viral envelope and on the HSV-infected cells or to internal proteins. They can also be utilized to determine the class (e.g., IgM, IgA, IgG, IgE) or subclass (e.g., IgG_1, IgG_2, etc.) of immunoglobulin. Of particular relevance has been the further understanding of the functional role of antibody. Neutralizing antibody (either alone or in the presence of complement) by definition lowers the titer of virus when virus is incubated with the antibody. This may be enhanced by agents such as rheumatoid factor (or other anti-antibodies) in

Division of Infectious Diseases, Department of Pediatrics, University of California Medical School, San Francisco, San Francisco General Hospital, 1001 Potrero Avenue, San Francisco, CA 94110, USA

Current Topics in Microbiology and Immunology, Vol. 179

addition to complement. Antibody alone apparently has no effect on the HSV-infected cells regarding viability. It indeed is probable that antibody may modulate cell surface antigens and alter both cell-viral interaction and interaction of components of the immune system with the cell in a potentially deleterious manner. In this regard the role of HSV glycoproteins that function as antibody receptors, i.e., glycoprotein E (gE) and gI, or complement receptors, i.e., gC are extensively discussed in the chapter by DUBIN et al., this volume. Unlike antibody alone, antibody and complement or antibody and various leukocytes are able to destroy viral-infected cells. The later mechanism is termed antibody-dependent cellular cytotoxicity (ADCC).

Thus, when determining a potential role for antibody in HSV infection, it is now becoming possible to discuss antibody to specific portions of virus, antibody of defined class, antibody of particular functional characteristics, and antibody to defined viral regions or epitopes. The advent of monoclonal antibody technology has very clearly demonstrated that a particular antibody may have an activity in one assay (e.g., ELISA) yet not in another (e.g., ADCC or neutralization).

2 Passive Protection Against HSV Infection by Antibody—Exogenous Administration

The clearest evidence for the role of antibody regarding the immune response to any pathogen is successful protection against infection or illness by passive administration of exogenous antibody. To my knowledge, there is no reputable study in humans in which exogenously administered antibody has been shown to prevent or ameliorate any type of HSV infection. There is quite substantial evidence that high titered antisera can protect immunodeficient human hosts against related human herpesviruses such as varicella zoster (BRUNELL et al. 1969; GERSHON et al. 1974; GEISER et al. 1975; ORENSTEIN et al. 1981) and cytomegalovirus (YAEGER et al. 1981; WINSTON et al. 1987; SNYDMAN 1990).

Contrary to the lack of evidence in humans, there is an abundance of evidence that antibody can protect animals against HSV infection. Antibody seems to play a role in prevention of severe infection (DIX et al. 1981), prevention of neuronal spread (MCKENDALL et al. 1970; OPENSHAW et al. 1979; SEKIZAWA et al. 1980), protection against recurrent disease (SIMMONS and NASH 1985), and maintenance of latency (STEVENS and COOK 1974). In a mouse model whereby initial passive antibody protection eventually resulted in an antibody-negative animal with latent virus, antibody was shown to be unnecessary for maintenance of latency (SEKIZAWA et al. 1980). Studies with recombinant HSV glycoprotein vaccines in mice have linked protection from infection to levels of neutralizing antibody (BLACKLAWS et al. 1987, 1990; WILEY et al. 1988). Utilization of monoclonal antibodies has more recently linked antibody protection to both neutralization and ADCC antibody activity against HSV gB, gC, and gD (MESTER et al. 1991; KOHL et al. 1990; KOHL 1991). These data will not be extensively reviewed. In the neonatal mouse model, relatively large doses of antibody are necessary to protect mice against small challenge doses of virus (KILBOURNE and HORSFALL 1951; LUYET et al. 1975; BARON et al. 1976; STRULOVITCH et al. 1979). Antibody must be adminstered before or shortly after (usually within 12 h of) HSV infection (GEORGIADES et al. 1982).

In the natural setting of murine transfer of antibody from mother to neonate, antibody is partially protective. Bodies born to nonimmune mothers but suckled by immune mothers were protected against a small challenge dose of HSV (HAYASHI et al. 1983; KOHL and LOO 1984). Proof of the role of antibody, and not colostrum, milk cells, cytokines, or other factors, was provided by experiments in which passive immunization of maternal nursing mice with exogenous antiviral antibody conferred protection against HSV infection to their suckling neonates (KOHL and LOO 1984). Thus in animals, but not in humans, there is clear evidence that antibody can protect against HSV, even in the immunodeficient neonate.

The issue becomes more complicated when one attempts protection against high doses of virus in the immunocompromised host. Thus RAGER-ZISMAN and ALLISON (1976) demonstrated that, in the immunocompromised adult mouse, a combination of antiviral antibody and spleen cells from nonimmune mice resulted in optimal antibody-mediated protection. Similarly, a series of work by OAKES' group demonstrated an antithymocyte serum or radiation-sensitive element that was critical in antibody-mediated protection in mice (OAKES et al. 1980; OAKES and ROSEMOND-HORNBEACK 1978; DAVIS et al. 1979). Supporting the importance of a cellular component is work demonstrating that intact antibody, but not $F(ab^1)_2$ fragments, were protective (OAKES and LAUSCH 1981; HAYASHIDA et al. 1982; MCKENDALL 1985; MESTER et al. 1991). This implies a role for the Fc portion of antibody, i.e., that part which interacts with the leukocyte Fc receptor to mediate mechanisms such as ADCC. Complement did not appear necessary (HAYASHIDA et al. 1982; MCKENDALL 1985). We have subsequently demonstrated, in a model involving a high viral dose challenge in neonatal mice, that antiviral antibody alone was not protective. When antibody was combined with leukocytes (either lymphocytes or macrophages) from adult humans, the mice were protected (KOHL and LOO 1982). Leukocytes from humans with ADCC effector cell defects, such as neonates (SHORE et al. 1977b; KOHL et al. 1984), or patients with cell surface CD11, CD18 deficiency (leukocyte adhesion defect) (KOHL et al. 1986) were not able to mediate this protection (KOHL et al. 1981, 1986). Most recently, utilizing a battery of polyclonal antisera to synthetic peptides of HSV gD and monoclonal antibodies (Mabs) to HSV gB, we have demonstrated that protection against a high challenge dose of virus was associated with the antisera or Mab ability to mediate ADCC in vitro. Protection required the combination of human leukocytes plus antibody in vivo. In contrast, a low challenge dose of virus was protected by the above mentioned antibodies and associated with antiviral neutralization (KOHL et al. 1990). These experiments are supported by a series of studies utilizing monoclonal antibodies, and, when tested, associating protection against HSV infection with ADCC activity of the antibody (BALANCHANDRAN et al. 1982; RECTOR et al. 1982, 1984; KINO et al. 1985; MESTER et al. 1991).

Thus, in animal models there is evidence that antibody is protective. In particular, neutralization antibody may be associated with protection against small doses of virus, while ADCC activity, in the presence of competent effector cells (either endogenous or adoptively transferred), was associated with protection against high challenge doses of virus.

3 Passive Protection by Transplacental Antibody Transfer in Human Neonates

The only human situation in which passive transfer of antibody often occurs, and might be analyzed for its effects alone on HSV infection, is in the neonate. The human neonate, unlike the mouse, receives essentially all passive antibody by placental transfer prior to delivery and does not absorb antibody systemically from milk or colostrum to any extent, except perhaps in the most premature neonates who may be born prior to "gut closure" (OGRA et al. 1977). Antibody to HSV is efficiently transported from the mother to the term infant (KOHL et al. 1978; SMITH and HANNA 1974; SHORE et al. 1976; SULLENDER et al. 1988). There have been several studies of human neonates with HSV infection and the possible effect of anti-HSV antibody. These are confounded by the problem that the infants have been studied upon or shortly after presentation of their illness, occurring from days to weeks after birth (Table 1). Thus, the antibody levels present in the neonates may be a reflection of declining transplacental passive antibody or rising neonatal endogenously produced antibody. Similarly, examining the levels of maternal antibody at the time of the infant's presentation may be misleading in the woman who had primary infection close to delivery. Such a woman will have low levels of antibody to transfer to the neonate yet, due to active antibody production, may have high levels when tested upon presentation of her sick neonate days later.

It is clear that the attack rate of neonatal herpes is an order of magnitude higher (approximately 30%–50%) in the neonate born vaginally to the mother with primary infection (NAHMIAS et al. 1971) than with recurrent infection (1%–3%) (PROBER et al. 1987; BROWN et al. 1991). One is tempted to ascribe this difference to the high levels of antibody in women with recurrent infection and the resultant similarly high levels in their neonates. Another major factor to consider is the longer duration and much higher titer of virus present in the vagina during primary infection (COREY et al. 1983). Thus, in addition to differences in antibody, there is in essence a different viral challenge dose to the infant born to the mother with primary HSV genital infection.

Table 1. Effect of transplacental HSV antibodies on human neonatal HSV infection

Study[a]	Number of babies	Time of neonate serum collection	Assay	Relationship of infection to antibodies
Whitley et al. 1980	35	First week from onset	Neutralization	Infection not related to antibody presence; all seropositive babies had titer > 40
Yeager et al. 1980	16	Upon presentation	Neutralization	Lower antibody levels in neonates with severe disease
Whitley et al. 1983	58	3 Days from onset	Neutralization	Infection not related to antibody presence
Sullender et al. 1987	47	Upon presentation	Neutralization	Babies with dissemination likely to lack antibody
Prober et al. 1987	40	Delivery	Neutralization	High titer of antibody associated with protection
Kahlon and Whitley 1988	25	Upon presentation	ELISA	Infection not related to antibody presence
Whitley et al. 1988	169	Upon presentation	Neutralization	Infection not related to antibody but neonates with CNS infection more likely to have antibody
Sullender et al. 1988	42	Upon presentation	ELISA	Low prevalence of HSV-2 gG in infected neonates
Kohl et al. 1989b	47	Upon presentation	Neutralization ADCC	Low levels of ADCC and neutralization antibodies associated with disseminated disease
Brown et al. 1991	47	Birth	Western blot	HSV-2 gG antibody associated with protection against neonatal HSV-2 infection

ELISA, enzyme-linked immunosorbent assay; gG, HSV glycoprotein G
[a] Many of the studies with similar authors may contain overlapping serum samples

With these provisos, it is of no great surprise that there have been discordances in the several studies of antibody and the human neonate. The early studies by the National Institutes of Health (NIH) Collaborative Antiviral Study Group, in which serum was collected within 1 week of disease onset, could discern no difference in severity of infection related to the absence or presence of antibody. A similar number of babies with antibody as without developed disseminated, localized CNS or skin only infection. Survival was not influenced by antibody status (WHITLEY et al. 1980). These results were confirmed by the same group in a cohort of 58 neonates (WHITLEY et al. 1983). When serum obtained from babies within 3 days of onset of illness was analyzed, equal numbers of babies with and without antibodies were classified in each category of severity (WHITLEY et al. 1983, 1988; KAHLON and WHITLEY 1988). Babies with encephalitis were more likely to have antibody than the others (WHITLEY et al. 1988).

In contrast to these studies have been those of the Stanford group and their collaborators. YEAGER et al. (1980) demonstrated that 55% of infants of mothers with primary HSV infection but 76% of those with recurrent disease had high titer (⩾ 1:40) neutralizing antibody. The titer of antibody in 5 neonates with mild disease (1:56 to type 1, 1:65 to type 2) at the onset of illness was higher than those of 11 neonates with severe infection (encephalitis or disseminated disease; 1:11 and 1:12 to HSV 1 and HSV 2, respectively). Exposed but uninfected neonates had the highest mean levels of antibody. In subsequent studies from this group, infants with disseminated HSV infection were shown to be more likely to lack neutralizing antibody in week 1 (73%) than those with encephalitis (11%) or mucocutaneous disease (35%) (SULLENDER et al. 1987). Infants exposed to HSV at delivery who remained asymptomatic had high concentrations of neutralizing antibody, while the 1 out of 40 infants who became infected had a low titer (PROBER et al. 1987). In an extension of these studies, the levels of both neutralizing and ADCC antibodies were analyzed. Those babies with ADCC levels above 10^{-3} did not have disseminated disease. In babies with lower levels, those with neutralizing antibody had less severe disease than those without. Babies exposed but uninfected had the highest ADCC antibody levels (KOHL et al. 1989b). These studies were confirmed in a smaller sample of subsets from the NIH group (KOHL et al. 1989b).

Recently, analysis of antibody to type-specific HSV-2 glycoproteins has become possible. Utilizing an ELISA for gG2, 92% of babies exposed to HSV-2 who remained asymptomatic had antibody to HSV-2 gG2. In contrast, only 12% of HSV-infected babies had antibody to gG2, suggesting a protective effect of antibody to this specific glycoprotein (SULLENDER et al. 1988). In a confirmatory study utilizing western blot analysis of serum, passive transfer of antibody to HSV-2 again appeared to protect the neonate from HSV-2 infection. In four cases of primary HSV-2 maternal infection, 1 (25%) of the neonates was infected. In 12 cases of HSV-2 initial infection, in which the mother was seropositive for antibodies to HSV-1 but lacked antibodies to HSV-2 gG, 4 (25%) of the neonates were infected. None of the neonates of the 31 women who were HSV-2 antibody-positive and gG-positive with HSV-2 reactivation at delivery were infected (BROWN et al. 1991). While both of these studies suggest that specific antibody to HSV-2 may be associated with protection, an alternate explanation is that the infected infant which lacked HSV-2 gG antibody was exposed to a primary or initial HSV-2 infection rather than a recurrence. This also would entail the effect of viral challenge dose to the neonate (being higher in primary or initial infection than in recurrence) and antibody-mediated passive protection.

In conclusion, primary maternal HSV infection at delivery is clearly associated with a higher rate of transmission of HSV to the neonate. Although controversial, recent evidence would suggest that this is due not only to viral dose, but also to a lack or low levels of functional (neutralization, ADCC) or specific (HSV-2 gG) antibodies in the neonate (Table 1). In addition, at least in several studies, low levels of functional antibody in the neonate (neutralization, ADCC) were associated with more severe (particularly disseminated) disease, while high levels of antibody were associated with local (particularly CNS) disease. The later finding of high levels of antibody in neonates with encephalitis and the usually late onset of this manifestation (more likely in the second week of life as compared to disseminated and mucocutaneous illness which typically occurs in the first week of life) have led to an alternate hypothesis. I have suggested that at least a portion of neonatal HSV encephalitis, as in adults, is due to reactivation of virus. In this scenario, virus may infect the neonate early with little or no symptoms (possibly due to high levels of antibody) but then, as in most cases of adult encephalitis, recur as symptomatic illness (KOHL 1990).

The question of the role of passive antibody protection against neonatal HSV infection is not merely an academic issue. Current preparations of immune globulin for intravenous use can be administered to the neonate in high doses (500–750 mg/kg) and provide adult levels of antibody. Unfortunately, in at least one published study, the levels of anti-HSV ADCC antibody provided by a commercial preparation were not in the range associated with protection from disseminated illness (KOHL et al. 1989a). Our laboratory has also noted similar results utilizing a second commercial preparation. Thus, as with other herpes virus infections, it is likely that, for antibody preparations to have an impact on neonatal HSV infection, hyperimmune globulin with high levels of specific and functional activity will be needed.

4 Severe HSV Infection in Humans with Antibody Deficiency Syndromes

The "experiments of nature" that have resulted in various immunodeficiency syndromes in humans have yielded important insights into the role of components of the immune response to types of pathogens. If antibody were a critically necessary component of the immune response to HSV infection, one would expect unusually severe cases of HSV infection in agammaglobulinemic patients. There are several reports of this occurrence. In the first of these, two children with X-linked (Bruton's) agammaglobulinemia were reported (LINNENMANN et al. 1973). These children both died of HSV encephalitis (one had uncomplicated "fever blisters" at some previous time). In one child an enterovirus was also recovered from the brain; the other child had received a live measles vaccination in the preceding week. In an intriguing report, a child with X-linked hypogammaglobulinemia with chronic cutaneous HSV infection was described (OLSON and HALL 1987). As in the previous report (LINNEMANN et al. 1973), the child had a chronic enteroviral CNS infection and was documented to have relatively few T4 cells (31.7%), low helper/suppressor ratio (0.80), and decreased response to concanavalin A. Of interest is the finding that treatment of the enteroviral encephalitis with high doses of intravenous gammaglobulin (500 mg/kg every 2 weeks) caused a 30% reduction in the size of the chronic HSV infection. Complete resolution occurred with antiviral therapy. While these cases of severe HSV occurred in children with pure B cell defects, the concomitant viral infection or live virus immunization caused the original authors and this writer to believe that these infections may have been due to a viral-mediated depression of cellular immunity and not just an antibody deficiency.

Common variable immunodeficiency is the classic non-X-linked hypogammaglobulinemia. In a series of eight children with this syndrome, two were reported to have generalized infection with HSV (CONLEY et al. 1986). This syndrome is heterogeneous and appears to involve defects in both T and B cell function.

In most reports of immunodeficiency patients, it is those with either T cell or combined defects that have had severe HSV infection (reviewed in KOHL 1987). Indeed, high levels of anti-HSV antibody have been associated with a greater likelihood of viral shedding after immunosuppression (PASS et al. 1979). Similarly, higher levels of antibody were found in humans susceptible to HSV recurrences than in unaffected seropositive controls by complement fixation (WILTON et al. 1972), indirect hemagglutination (THONG et al. 1975), and neutralization (LOPEZ and O'REILLY 1977). An interesting recent report describes preliminary studies showing a slight decline in ADCC antibody activity at the time of genital HSV recurrence (OH et al. 1989). Antibody can be shown to interfere with the protective activity of neutralizing antibody (ASHE and NOTKINS 1967). It is most investigators' belief that the high levels of antibody seen in individuals with frequently recurring HSV are due to intense antigenic stimulation and not that the high levels facilitate recurrence.

In a recent study of patients with renal transplants or leukemia, the rate of reactivation of HSV and clinical symptom were compared (GREENBERG et al. 1987). While both groups had similar rates of viral reactivation (47% vs 50%) only 32% of transplant patients who had reactivated HSV developed lesions, while 100% of the leukemic patients did. In immunologic studies, the major difference between these two groups was in the low ADCC effector activity of the leukemic patients. The latter would be expected to have ADCC mediating antibody, and perhaps the poor effector cell function contributed to severe HSV infection. ADCC effector cell defects have been similarly demonstrated in patients with extensive burns at a time in which they are most susceptible to severe HSV infection (KOHL and ERICSSON 1982). A fascinating report described a girl with an unusual susceptibility to severe herpes infections (including HSV) and absence of natural killer (NK) cells (BIRON et al. 1989). Since the NK cell is the major peripheral blood cell which mediates ADCC (SHORE et al. 1977a), it is no surprise that this patient also lacked ADCC functional activity.

Thus, antibody deficiency alone does not appear to be associated with severe HSV infection, and frequent or severe HSV infection can occur in the normal and immunocompromised host in the presence of antibody. As in several animal models noted above, cells which interact with antibody to mediate ADCC, and thus facilitate antibody function, may be important in immunodeficient hosts and in the neonate.

5 The Effect of Acyclovir on the Antibody Response to HSV

It is clear that systemic acyclovir treatment is associated with a lessening of severity of first episode genital HSV infection and of HSV infection in the immunocompromised patient. Similarly, chronic suppressive acyclovir therapy is associated with a marked decrease in the rate of recurrent infection (reviewed in KOHL 1987). The effect of acyclovir therapy on the immune response and subsequent clinical course of HSV infection was initially felt to shed light on the importance of the antibody response to HSV.

Several groups have shown that therapy of the first episode of genital HSV infection with either oral or intravenous acyclovir was associated with a decreased and delayed antibody response (ASHLEY and COREY 1984; BERNSTEIN et al. 1984; ASHLEY et al. 1988). This was seen in response to specific viral components including 11g80, gD, and Vp66 (ASHLEY and COREY 1984); gD and gE (BERNSTEIN et al. 1984); and gB, gC, gE, Vp16, gD, and p45 (ASHLEY et al. 1988). A decreased production of type-specific neutralizing antibody was also detected (BERNSTEIN et al. 1984). Of particular importance has been an association of the lower and delayed antibody response in the acyclovir recipients with a more severe first recurrence (ASHLEY and COREY 1984; ASHLEY et al. 1988). Thus it appeared that the role of antibody was important in determining the severity of recurrent genital infection.

These initial impressions have been confounded by subsequent studies of the cell-mediated immune response in normal and immunocompromised patients receiving acyclovir. In the normal patient treated with acyclovir for the first episode of genital HSV infection, the lymphocyte transformation response to HSV was much lower and took longer to peak than in control patients (LAFFERTY et al. 1984). Similarly, in bone marrow transplant patients treated with acyclovir for prophylaxis (WADE et al. 1984b; LJUNGMAN et al. 1986) or therapy of HSV infection (WADE et al. 1984a), the lymphocyte response was lower than in placebo-treated patients.

Thus, while short-term acyclovir therapy is associated with a blunted initial antibody response and more severe recurrent disease, the effect of acyclovir on the cellular response as well makes it impossible to use acyclovir treatment in order to assess the role of antibody to HSV in disease severity.

The picture is even more clouded in the case of chronic suppressive acyclovir therapy. Several groups have shown that long-term acyclovir chemoprophylaxis of genital infection was associated with a marked reduction in HSV clinical reactivation and in antibody levels to HSV (GOLD et al. 1988; ERLICH et al. 1988). A recent report has failed to detect a decline in neutralizing antibody levels and instead demonstrated a rise in lymphocyte transformation response to HSV antigen during chronic acyclovir therapy (FRENKEL et al. 1989). High levels of antibody to gB have been associated with more severe recurrence after cessation of therapy (GOLD et al. 1988).

In conclusion, it appears that acyclovir acute systemic therapy is associated with a blunting of both the humoral and cellular immune response to HSV infection and a more severe first recurrence. Due to the effect on both arms of the immune system, it is impossible to use these observations in attempting to ascertain the role of antibody in the human response of HSV infection. There probably is also a decreased antibody response during chronic therapy, at a time of decreased viral activity. Whether the mechanism of the altered immune response is due to a diminution of viral proliferation, resulting in less antigen stimulation, or to a more direct effect of acyclovir on the immune system remains to be clarified (STEELE et al. 1980; LEVIN et al. 1980).

6 Conclusions

In animal models, it is clear that exogenous antibody prevents or modifies many aspects of HSV infection. These experiments point to both neutralizing and ADCC functional antibody as being important, and are beginning to define epitopes, particularly on gB and gD, of special interest. In humans, the studies involving transplacental antibody in neonates remain controversial, but appear to indicate a role of antibody in determination of severity of infection. While high titer passive antibody has been shown to prevent related herpesvirus infections in immunocompromised humans, these studies remain to be done regarding HSV. The role of HSV antibody is confounded by alternate defense mechanisms and the possible role of the cellular arm of the immune system in optimizing the function of antibody (as in ADCC activity). Defining the role of antibody in humans regarding HSV infection is of importance since it may be administered to high risk hosts. Similarly, as vaccine development advances, the necessary immune responses mediating protection are critical. If antibody is involved, then functional type and epitope specificity assume major importance.

Acknowledgment. Support by National Institutes of Health Program Project Grant HD13021.

References

Ashe WK, Notkins AL (1967) Kinetics of sensitization of herpes simplex virus and its relationship to the reduction in the neutralization rate constant. Virology 33: 613–617
Ashley RL, Corey L (1984) Effect of acyclovir treatment of primary genital herpes on the antibody response to herpes simplex virus. J Clin Invest 73: 681–688
Ashley R, Mack R, Critchlow C, Shurtleff M, Corey L (1988) Differential effect of systemic acyclovir treatment of genital HSV-2 infections on antibody responses to individual HSV-2 proteins. J Med Virol 24: 309–320
Balachandran N, Bacchetti S, Rawls WE (1982) Protection against lethal challenge of BalB/C mice by passive transfer of monoclonal antibodies to five glycoproteins of herpes simplex virus type 2. Infect Immun 37: 1132–1137
Baron S, Worthingon MG, Williams J, Gaines JW (1976) Post exposure serum prophylaxis of neonatal herpes simplex virus infection of mice. Nature 261: 505–506
Bernstein DI, Lovett MA, Bryson YJ (1984) The effects of acyclovir on antibody response to herpes simplex virus in primary genital herpetic infections. J Infect Dis 150: 7–13
Biron CA, Bryon KS, Sullivan JL (1989) Severe herpes virus infections in an adolescent without natural killer cells. N Eng J Med 320: 1731–1735
Blacklaws BA, Krishna S, Minson AC, Nash AA (1990) Immunogenicity of herpes simplex virus type 2 glycoproteins expressed in vaccinia virus recombinants. Virology 177: 727–736
Blacklaws BA, Nash AA, Darby G (1987) Specificity of the immune response of mice to herpes simplex virus glycoprotein B and D constituitively expressed on L cell lines. J Gen Virol 68: 1103–1114
Brown ZA, Benedetti J, Ashley R, Burchett S, Selke S, Berry S, Vontver LA, Corey L (1991) Neonatal herpes simplex virus infection in relation to asymptomatic maternal infection at the time of labor Engl J Med 324: 1247–1252
Brunell PA, Ross A, Miller LH, Kuo B (1969) Prevention of varicella by zoster immunoglobulin. N Engl J Med 280: 1191–1194
Conley ME, Park CL, Douglas SD (1986) Childhood common variable immunodeficiency with autoimmune disease, J Pediatr 108: 915–922
Corey L, Adams HG, Brown ZA, Holmes KK (1983) Genital herpes simplex virus infections, clinical manifestations, course, and complications. Ann Intern Med 98: 958–972
Davis WB, Taylor JA, Oakes JE (1979) Ocular infection with herpes simplex virus type 1: prevention of acute herpetic encephalitis by systemic administration of virus specific antibody. J Infect Dis 140: 534–540
Dix RD, Pereira L, Baringer JR (1981) Use of monoclonal antibody directed against herpes simplex virus glycoproteins to protect mice against acute virus-induced neurological disease. Infect Immun 34: 192–199
Erlich KS, Hauer L, Mills J (1988) Effects of long-term acyclovir chemosuppression on serum IgG antibody to herpes simplex virus. J Med Virol 26: 33–39
Frenkel L, Pineda E, Garratty E, Fall H, Dilon M, Bryson Y (1989) A perspective study of the effects of acyclovir treatment on the HSV-2 lymphoproliferative response of persons with frequently recurring HSV-2 genital infection. J Infect Dis 159: 845–850
Geiser CF, Bishop Y, Myers M, Jaffe N, Yankee R (1975) Prophylaxis of varicella in children with neoplastic disease: comparative results with zoster immune plasma and gamma globulin. Cancer 35: 1027–1030
Georgiades JA, Montgomery J, Hughes TK, Jensen D, Baron S (1982) Determinants of protection by human immune globulin against experimental herpes neonatorum. Proc Soc Exp Biol Med 170: 291–297
Gershon AA, Steinberg S, Brunell PA (1974) Zoster immune globulin. A further assessment. N Engl J Med 290: 243–245
Gold D, Ashley R, Solberg G, Abbo H, Corey L (1988) Chronic-dose acyclovir to suppress frequently recurring genital herpes simplex virus infection: effect on antibody response to herpes simplex virus type 2 proteins. J Infect Dis 158: 1227–1234
Greenberg MS, Friedman H, Cohen SG, Oh SH, Laster L, Starr S (1987) A comparative study of herpes simplex infections in renal transplant and leukemic patients. J Infect Dis 156: 280–287
Hayashi Y, Wada T, Mori R (1983) Protection of newborn mice against herpes simplex virus infection by prenatal and postnatal transmission of antibody. J Gen Virol 64: 1007–1012

Hayashida I, Nagafuchi S, Hayashi Y, Kino Y, Mori R, Oda H, Ohtomo N, Tashiro A (1982) Mechanism of antibody-mediated protection against herpes simplex virus infection in athymic nude mice: Requirement of Fc portion of antibody. Microbiol Immunol 26: 497–509

Kahlon J, Whitley RJ (1988) Antibody response of the newborn after herpes simplex virus infection. J Infect Dis 158: 925–933

Kilbourne EM, Horsfall FL Jr (1951) Studies of herpes simplex virus in newborn mice. J Immunol 67: 321–329

Kino Y, Eto T, Ohtomo N, Hayashi Y, Yamamoto M, Mori R (1985) Passive immunization of mice with monoclonal antibodies to glycoprotein gB of herpes simplex virus. Microbiol Immunol 29: 143–149

Kohl S (1987) Postnatal herpes simplex virus infections. In: Feigin RD, Cherry JD (eds) Textbook of pediatric infectious diseases, 2nd. Saunders, Philadelphia, pp 1577–1601

Kohl S (1990) A hypothesis on the pathophysiology of neonatal herpes simplex virus encephalitis: clinical recurrence after asymptomatic primary infection. Pediatr Infect Dis J 9: 307–308

Kohl S (1991) Role of antibody-dependent cellular cytotoxicity in defense against herpes simplex virus infection. Rev Infect Dis 13: 108–114

Kohl S, Erisson CD (1982) Cellular cytotoxocity to herpes simplex virus-infected cells of leukocytes from patients with serious burns. Clin Immunol Immunopathol 24: 171–178

Kohl S, Loo LS (1982) Protection of neonatal mice against herpes simplex virus infection. Probable in vivo antibody-dependent cellular cytoxocity. J Immunol 129: 370–376

Kohl S, Loo LS (1984) The relative role of transplacental and milk immune transfer in protection against lethal neonatal herpes simplex virus infection in mice. J Infect Dis 149: 38–42

Kohl S, Shaban SS, Starr SE, Wood PA, Nahmias AJ (1978) Human neonatal and maternal monocyte-macrophage and lymphocyte-mediated antibody-dependent cytotoxicity to cells infected with herpes simplex. J Pediatr 93: 206–210

Kohl S, Loo LS, Pickering LK (1981) Protection of neonatal mice against herpes simplex viral infection by human antibody and leukocytes from adult but not neonatal humans. J Immunol 127: 1273–1275

Kohl S, Loo LS, Gonick B (1984) Analysis in human neonates of defective antibody-dependent cellular cytotoxicity and natural killer cytotoxicity to herpes simplex virus-infected cells. J Infect Dis 150: 14–19

Kohl S, Loo LS, Schmalstieg FS, Anderson DC (1986) The genetic deficiency of leukocyte surface glycoprotein Mac-1, LFA-1, p150, 95 in humans in associated with defective antibody-dependent cellular cytotoxicity in vitro and defective protection against herpes simplex virus infection in vivo. J Immunol 137: 1688–1694

Kohl S, Loo LS, Rench MA, Noya FSD, Feldman S, Baker CJ (1989a) Effect of intravenously administered immune globulin on functional antibody to herpes simplex virus in low birth weight neonates. J Pediatr 115: 135–139

Kohl S, West MS, Prober CG, Loo LS, Sullander W, Arvin AM (1989b) Neonatal antibody-dependent cellular cytotoxicity antibody levels are associated with the clinical presentation of neonatal herpes simplex virus infection. J Infect Dis 160: 770–776

Kohl S, Strynadka NCJ, Hodges RS, Pereira L (1990) Analysis of the role of antibody-dependent cellular cytotoxicity antibody activity in murine neonatal herpes simplex virus infection with antibodies to synthetic peptides of glycoprotein D and monoclonal antibodies to glycoprotein B. J Clin Invest 86: 273–278

Lafferty WE, Brewer LA, Corey L (1984) Alteration of lymphocyte transformation response to herpes simplex virus by acyclovir therapy. Antimicrob Agents Chemother 26: 887–891

Levin MJ, Leary PL, Arbeit RD (1980) Effect of acyclovir on the proliferation of human fibroblasts and peripheral blood mononuclear cells. Antimicrob Agents Chemother 17: 947–953

Linneman CC Jr, May DB, Schubert WK, Caraway CT, Schiff GM (1973) Fatal viral encephalitis in children with X-linked hypogammaglobulinemia. Am J Dis Child 126: 100–103

Ljungman P, Wilizek H, Gahrton G, Gustavsson A, Lundgren G, Lonnqvist B, Ringden O, Wahren B (1986) Long-term acyclovir prophylaxis is bone marrow transplant recipients and lymphocyte proliferation responses to herpes virus antigens in vitro. Bone Marrow Transplant 1: 185–192

Lopez C, O'Reilly RJ (1977) Cell-mediated immune responses in recurrent herpes virus infections. J Immunol 118: 895–902

Luyet F, Samra D, Soneji A, Marks MI (1975) Passive immunization in experimental herpes virus hominis infection of newborn mice. Infect Immun 12: 1258–1261

McKendall RR (1985) IgG-mediated viral clearance in experimental infection with herpes simplex virus type 1: role for neutralization and Fc-dependent functions but not Cl cytolysis and C5 chemotaxis. J Infect Dis 151: 464–470
McKendall RR, Klassen T, Baringer JR (1979) Host defenses in herpes simplex infections of the nervous system: Effect of antibody on disease and viral spread. Infect Immun 23: 305–311
Mester JC, Glorioso JC, Rouse BT (1991) Protection against zosteriform spread of herpes simplex virus by monoclonal antibodies. J Infect Dis 163: 263–269
Nahmias AJ, Josey WE, Naib ZM, Freeman MG, Fernandez RJ, Wheeler JH (1971) Perinatal risk associated with maternal genital herpes simplex virus infection. Am J Obstet Gynecol 110: 825–837
Oakes JE, Lausch RN (1981) Role of Fc fragment in antibody-mediated recovery from ocular and subcutaneous herpes simplex virus infections. Infect Immun 33: 109–114
Oakes JE, Rosemond-Hornbeak H (1978) Antibody-mediated recovery from subcutaneous herpes simplex virus type 2 infection. Infect Immun 21: 489–495
Oakes JE, Davis WB, Taylor JA, Weppner WA (1980) Lymphocyte reactivity contributes to protection conferred by specific antibody passively transferred to herpes simplex virus-infected mice. Infect Immun 29: 642–649
Ogra SS, Weintraub D, Ogra PL (1977) Immunologic aspects of human colostrum and milk. III. Fate and adsorption of cellular and soluble components in the gastrointestinal tract of the newborn. J Immunol 119: 245–248
Oh SH, Douglas JM, Corey L, Kohl S (1989) Kinetics of the humoral immune response measured by antibody-dependent cell-mediated cytotoxicity and neutralization assays in genital herpes simplex infections. J Infect Dis 159: 328–330
Olson NY, Hall JC (1987) Chronic cutaneous herpes simplex and X-linked hypogammaglobulinemia. Pediatr Dermatol 4: 225–228
Openshaw H, Asher LVS, Wohlenberg C, Sekizawa T, Notkins AL (1979) Acute and latent infection of sensory ganglia with herpes simplex virus: immune control and virus reactivation. J Gen Virol 44: 205–215
Orenstein WA, Heymann DL, Ellis RJ, Rosenberg RL, Nakano J, Halsey NA, Overturf GD, Hayden GF, Witte JS (1981) Prophylaxis of varicella in high risk children: dose-response effect of zoster immune globulin. J Pediatr 98: 368–373
Pass RF, Whitley RS, Whelchel JD, Diethelm AG, Reynolds DW, Alford CA (1979) Identification of patients with increased risk of infection with herpes simplex virus after renal transplantation. J Infect Dis 140: 487–492
Prober CG, Sullender WM, Yasukawa LL, Au DS, Yeager AS, Arvin AM (1987) Low risk of herpes simplex virus infection in neonates exposed to the virus at the time of vaginal delivery to mothers with recurrent genital herpes simplex virus infection. N Engl J Med 316: 240–244
Rager-Zisman B, Allison AC (1976) Mechanism of immunologic resistance to herpes simplex virus 1 (HSV-1) infection. J Immunol 116: 35–40
Rector JT, Lausch RN, Oakes JE (1982) Use of monoclonal antibodies for analysis of antibody-dependent immunity to ocular herpes simplex virus type 1 infection. Infect Immun 38: 168–174
Rector JT, Lausch RN, Oakes JE (1984) Identification of infected cell-specific monoclonal antibodies, and their role in host resistance to ocular herpes simplex virus type 1 infection. J Gen Virol 65: 657–666
Sekizawa T, Openshaw H, Wohlenberg C, Notkins AL (1980) Latency of herpes simplex virus in absence of neutralizing antibody: Model for reactivation. Science 210: 1026–1028
Shore SL, Black CM, Melewicz FM, Wood PA, Nahmias AJ (1976) Antibody-dependent cell mediated cytotoxicity to target cells infected with type 1 and type 2 herpes simplex virus. J Immunol 116: 194–201
Shore SL, Melewicz FM, Gordon DS (1977a) The mononuclear cell in human blood which mediates antibody dependent cellular cytotoxicity to virus-infected target cells: I. Identification of the population of effector cells. J Immunol 118: 558–566
Shore SL, Milgrom H, Wood P, Nahmias AJ (1977b) Neonatal function of antibody-dependent cell-mediated cytotoxicity to target cells infected with herpes simplex virus. Pediatrics 59: 22–28
Simmons A, Nash AA (1985) Role of antibody in primary and recurrent herpes simplex virus infection. J Virol 53: 944–948
Smith RN, Hanna L (1974) Herpes virus infections in pregnancy: a comparison of neutralizing antibody titers in mothers and their infants. Am J Obstet Gynecol 119: 314–318

Snydman DR (1990) Cytomegalovirus immunoglobulins in the prevention and treatment of cytomegalovirus disease. Rev Infect Dis 12: S839–S848
Steele RW, Marmer DJ, Keeney RE (1980) Comparative in vitro immunotoxicology of acyclovir and other antiviral agents. Infect Immun 28: 957–962
Stevens JG, Cook ML (1974) Maintenance of latent herpetic infection: an apparent role for anti-viral IgG. J Immunol 113: 1685–1693
Strulovitch C, Marks MI, Soneji A, Goldberg S (1979) Immunotherapy and drugs in neonatal disseminated herpes simplex virus type 2: a mouse model. J Antimicrob Chemother 5: 437–446
Sullender WM, Miller JL, Yasukawa LL, Bradley JS, Black SB, Yeager AS, Arvin AM (1987) Humoral and cell-mediated immunity in neonates with herpes simplex virus infection. J Infect Dis 155: 28–37
Sullender WM, Yasukawa LL, Schwartz M, Pereira L, Hensleigh PA, Prober CG, Arvin AM (1988) Type-specific antibodies to herpes simplex virus type 2 (HSV-2) glycoprotein G in pregnant women, infants exposed to maternal HSV-2 infection at delivery, and infants with neonatal herpes. J Infect Dis 157: 164–171
Thong YH, Vincent MM, Hensen SA, Fuccillo DA, Rola-Pleszczynski M, Bellanti JA (1975) Depressed specific cell-mediated immunity to herpes simplex virus type 1 in patients with recurrent herpes labialis. Infect Immun 12: 76–80
Wade JC, Day LM, Crowley JJ, Meyers JD (1984a) Recurrent infection with herpes simplex virus after marrow transplantation: role of the specific immune response and acyclovir treatment. J Infect Dis 149: 750–756
Wade JC, Newton B, Flournoy N, Meyers JD (1984b) Oral acyclovir for prevention of herpes simplex virus reactivation after marrow transplantation. Ann Intern Med 100: 823–828
Whitley RJ, Nahmias AJ, Visintine AM, Fleming CL, Alford CA (1980) The natural history of herpes simplex virus infection of mother and newborn. Pediatrics 66: 489–494
Whitley RJ, Yeager A, Kartus B, Bryson Y, Connor JD, Alford CA, Nahmias A, Soong S-J (1983) Neonatal herpes simplex virus infection: follow-up evaluation of Vidarabine therapy. Pediatrics 72: 778–785
Whitley RJ, Corey L, Arvin A, Lakeman FD, Sumaya CV, Wright PF, Dunkle LM, Steele RW, Soong SJ, Nahmias AJ, Alford CA, Powell DA, San Joquin V (NIAID Collaborative Antiviral Study Group) (1988). Changing presentation of herpes simplex virus infection in neonates. J Infect Dis 158: 109–116
Willey DE, Cantin EM, Hill LR, Moss B, Notkins AL, Openshaw H (1988) Herpes simplex virus type 1—vaccinia recombinant expressing glycoprotein B: protection from acute and latent infection. J Infect Dis 158: 1382–1386
Wilton JMA, Ivanyi L, Lehner T (1972) Cell-mediated immunity in herpes virus hominis infections. Br Med J 1: 723–726
Winston DJ, How G, Lin CH, Bartoni K, Budinger MD, Gale RP, Champlin RE (1987) Intravenous immune globulin for prevention of cytomegalovirus infection and interstitial penumonia after bone marrow transplantation. Ann Intern Med 106: 12–18
Yeager AS, Arvin AM, Urbani LJ, Kemp JA (1980) Relationship of antibody to outcome in neonatal herpes simplex virus infections. Infect Immun 29: 532–538
Yeager AS, Grumet FC, Hafleigh EB, Arvin AM, Bradley JE, Prober CG (1981) Prevention of transfusion-acquired cytomegalovirus infections in newborn infants. J Pediatr 98: 281–287

Macrophages and Other Nonspecific Defenses: Role in Modulating Resistance Against Herpes Simplex Virus

L. WU and P. S. MORAHAN

1 Introduction

Nonspecific host resistance is provided by effector cell systems, including the mononuclear phagocytes, i.e., circulating monocytes, and tissue macrophages (MOs), natural killer (NK) cells, and polymorphonuclear granulocytes (PMNs).

Department of Microbiology and Immunology, The Medical College of Pennsylvania, 3300 Henry Avenue, Philadelphia, PA 19129, USA

Current Topics in Microbiology and Immunology, Vol. 179

These cells, together with cytokines, play various roles at different stages in the pathogenesis of HSV infection (MORAHAN et al. 1985; MOGENSEN 1979). Immunomodulation of nonspecific resistance is one useful approach to control HSV infection (MORAHAN and MURASKO 1989). It has been effective in reducing mortality against HSV infections in immunosuppressed animals (MORAHAN and PINTO 1991) and may be especially useful in combination with chemotherapy (CONNELL et al. 1985).

Elucidation of the natural resistance mechanisms against HSV has focused on three experimental approaches: (1) selective depletion of effector cell populations using monoclonal antibodies, ^{89}Sr, and cell toxins; (2) adoptive transfer of selected effector cell populations, cloned cells, or mediators produced by cells; and (3) kinetic studies correlating increases in certain mediators or effector cell populations and functions with critical stages in viral pathogenesis.

Molecular biotechnological developments are beginning to be applied to studying nonspecific antiviral resistance mechanisms. Transgenic mice have been used to establish the viral DNA sequences that govern the MO tissue tropism of visna virus (SMALL et al. 1989). Genetically engineered virus mutants have been used to define the MO tropism of herpes simplex virus (HSV) (BEN-HUR et al. 1988) and the genes required for establishment of latency and neuroinvasiveness (IZUMI and STEVENS 1990). Transfection techniques have been developed to transfect mouse resident peritoneal MOs to study the activities of viral and cellular promoters in primary nonreplicating MOs (WU et al. 1990). The application of these molecular experimental approaches, in combination with kinetic correlation, cell depletion, and transfer studies, will lead to a better understanding of the mechanisms of nonspecific resistance to HSV infections.

2 Critical Role of MOs in Natural Resistance to HSV Infection

The importance of MOs, in controlling virus infection has long been recognized in experimental animal systems. These cells monitor the main body compartments and establish a first line of defense following entry into target organs (JOHNSON 1964). Our understanding of MO antiviral mechanisms, the interaction of MOs with other nonspecific and specific aspects of the immune system, and the relevance of MO-mediated resistance for natural resistance or immunotherapy in HSV infections in humans is, however, still rudimentary. Two general types of MO-mediated resistance mechanisms have been defined. (1) The capability of MOs to inactivate extracellular virus, inhibit virus replication in surrounding cells which are permissive to virus infection, or destory infected cells is defined as *extrinsic resistance.* (2) The ability of the MO to inhibit virus growth inside the MO itself is defined as *intrinsic resistance* (MORAHAN et al. 1985). Both mechanisms can be mediated by MOs either alone, or in concert with other immune effector cells or molecules.

The most definitive evidence for the role of MOs in resistance to HSV has come from cell transfer studies. JOHNSON (1964) firmly established the importance of MOs in HSV-1 infection by showing that adoptive transfer of MOs from HSV-1-infected adult mice to naive young mice transferred protection from lethal HSV-1 infection. The most consistent results have been obtained with transfer of peritoneal cells from immunomodulator-treated adult mice to naive suckling mice (MORAHAN and MORSE 1979).

The exact mechanism of the resistance induced by transferred peritoneal MOs remains to be determined. It may be mediated by MOs alone, by extrinsic and/or intrinsic mechanisms, or in concert with other cell populations or factors (such as cytokines and antibody). Transferred resident peritoneal MOs have been recently reported to enter the circulation, localize, and interact with other cells through adhesion molecules (ROSEN and GORDON 1990). Now that more information is available on the heterogeneity of mononuclear phagocytes (MORAHAN et al. 1988), it would be helpful to transfer more highly purified and stably labeled cell populations.

Experimental evidence, in addition to that provided by cell transfer, also points to an important role for MOs. For instance, depletion of tissue MOs markedly decreases the resistance of mice to HSV infection (PINTO et al. 1991; MORAHAN and MORSE 1979). The effect is more marked if: (a) tissue MOs rather than circulating monocytes are depleted, (b) tissue MOs in the area around the initial sites of infection are destroyed shortly before or after infection, or (c) MOs are destroyed rather than functionally inhibited (MORAHAN et al. 1986; PINTO et al. 1991).

A recent molecular study has also pointed to a role for the MO. Active immediate early (IE) HSV-1 gene expression in MOs correlated closely with the outcome of infection with intratypic recombinants with or without the IE region (BEN-HUR et al. 1988). The investigators speculated that the IE gene region sequence was required for viral gene expression and was recognized by cellular factors in infected MOs. This was the first report providing molecular evidence for a correlation between viral gene expression in MOs and pathogenesis. Another recent molecular pathogenesis study used in situ hybridization and immunocytochemistry to show a close correlation between liver MOs and the different degrees of hepatitis produced by HSV-1 and HSV-2 in mice (SCHIRMACHER et al. 1989).

There is some evidence in humans for the importance of MOs in resistance. It has been generally observed that MOs or monocytes from neonates are less resistant to HSV infection than are those from adults (reviewed in MORAHAN and MORSE 1979). However, marked depression of circulating leukocytes, such as in cancer patients receiving chemotherapy, is not usually accompanied by HSV infection. A possible explanation would be maintenance of natural MO resistance in tissues, perhaps in combination with specific anti-HSV immune responses.

3 Extrinsic Resistance of MOs to HSV Infection

3.1 Modulation by MO Activation State or Age of the Animal

Mo activation increases MO extrinsic resistance to HSV (reviewed in MORAHAN et al. 1980). The diverse activation signals include treatment of MOs in vitro with lymphokines and interferons (IFNs), injection of mice with sterile inflammatory agents (e.g., thioglycollate, proteose peptone broths), immunomodulators (e.g., *C. parvum*, BCG), infection with HSV, and immunization with vaccinia virus.

Extrinsic antiviral activity is also clearly age related. Suckling and weanling mice generally cannot be protected by treatment with immunomodulators as effectively as can adult mice (MORAHAN et al. 1980). The ability of MOs to be activated, the increased level of antiviral activity on a per cell basis, and the increased numbers of MOs available in older as compared with younger mice all may contribute to the well documented age-related increase in resistance to HSV infection in vivo (reviewed in MORAHAN et al. 1985).

3.2 Potential Mechanisms

Extrinsic antiviral resistance of MOs may occur at different levels in the viral replicative cycle, and which mechanism predominates may depend on: (a) the MO activation or differentiation state, (b) the virus replication strategy and pathogenesis pattern, and (c) the properties of the particular permissive host cell.

3.2.1 Inhibition of Virus Spread

HSV can spread to nearby or distant cells by the extracellular route and to adjacent cells by interecellular bridges. Inhibition of direct spread between cells by leukocytes in the absence of antibody was reported by LODMELL et al. (1973). The effector cells were peritoneal MOs but they needed to be activated to inhibit replication or viral spread (HAYASHI et al. 1980; MORAHAN et al. 1980). The antiviral activity occurred early in infection and was largely independent of the antibody or immune status of the mice.

3.2.2 Cytotoxicity for Virus-Infected Cells

Neither peptone-elicited MOs nor *C. parvum*-activated MOs showed increased cytotoxicity against HSV-infected cells as compared with uninfected cells (HAYASHI et al. 1980; MORSE and MORAHAN 1981). This suggests early nonspecific inhibition of viral replication or spread, before sufficient viral replication enables infected cells to be recognized and destroyed by immune effector cells. In certain systems using lymphokine-activated MOs however, antiviral activity has been correlated with selective lysis of HSV-infected cells, leaving uninfected cells unharmed

(KOFF et al. 1983). These differences in antiviral mechanisms that occur with various types of MO activation and target cells have not yet been analyzed.

3.2.3 Inhibition of Virus Replication in Permissive Cells

Several features appear characteristic of extrinsic MO-mediated antiviral activity: (a) virus nonspecificity against DNA and RNA viruses; (b) species nonspecificity; (c) requirement for live MOs; (d) requirement for contact between the MO and the permissive cell; and (e) virus multiplicity dependency (MORAHAN et al. 1980). Research has been hampered by the fact that MO extrinsic antiviral activity is most apparent against multiple cycles of infection initiated at a low multiplicity of infection (analagous to the in vivo situation). These biological features point to an IFN-independent, but virus-nonspecific, inhibition of viral replication, perhaps secondary to effects of the MO on the metabolism of the permissive cell. Activated MOs can produce profound aberrations in synthesis of DNA, RNA, and protein in target cells (KAPLAN et al. 1978); these could certainly affect HSV replication, which requires many host enzymes.

The molecular basis for inhibition of HSV replication is unclear. Reported mechanisms that involve MO secretion of arginase, and not IFN, need additional experimental support (KOFF et al. 1983). The studies on arginase used media very low in arginine in order to demonstrate the effect, but the relevance of this system to the in vivo situation is unclear. The studies with IFN used crude antiserum against IFN-α/β and need to be repeated with highly specific antibodies. Also, there is little information about cytokines other than IFN that may be produced by antivirally active MOs.

Information is also lacking on the molecular basis of extrinsic resistance mediated by a pure population of MOs, MO clones, or cell lines (MORAHAN et al. 1985). LEIBSON et al. (1986) reported that human mononuclear cells inhibited plaque formation of HSV in human fibroblast monolayers, that the inhibition did not require T cells, and that the inhibition was reduced if the cells were pretreated with anti-Macl antibody. Since this antibody recognizes MOs, PMNs, and NK cells, the role of MOs alone is not clear.

4 Intrinsic Resistance of MOs to HSV Infection In Vitro

4.1 Potential Mechanisms

Through their phagocytic function, MOs may take up and digest virus particles and degrade viral genomes, thus preventing or delaying the spread of virus (MOGENSON 1979). Human peripheral blood monocytes may spontaneously take up 25% of a viral inoculum in the first 30 min, gradually increasing the amount over 2 h. The addition of opsonizing antibody plus complement markedly increased uptake (VAN STRIJP et al. 1989a). Rapid degradation of the internalized HSV proteins and DNA occurred independent of toxic oxygen metabolites (VAN STRIJP et al. 1990).

Other intrinsic MO-mediated antiviral resistance mechanisms may involve a block in productive infection that could theoretically occur at any stage of viral infection, including attachment of virus to receptors, viral uncoating, transcriptional and posttranscriptional regulation of virus gene expression, translational and posttranslational control of viral proteins, initiation of DNA replication, virus assembly, and release of infectious viral particles (WU et al. 1990). The marked intrinsic resistance that MOs exhibit to HSV infection probably involves the combination of the MO classic phagocytic and intracellular destruction processes, coupled to inhibition of viral replication.

4.1.1 Molecular Basis for the Block in HSV-1 Replication in Mouse Resident Peritoneal MOs

The interaction of human, mouse, and rabbit MOs with HSV-1 has received considerable attention since the seminal studies of JOHNSTON (1964); STEVENS and COOK (1970) established that resident peritoneal MOs from adult but not neonatal mice were nonpermissive to HSV-1 infection. Most of the research has been performed with resident peritoneal MOs from naive mice, which serve as a "gold standard" for comparative purposes (MORAHAN et al. 1989).

Mouse resident peritoneal MOs are virtually nonpermissive to HSV-1 replication, although HSV enters these cells as efficiently as it does fully permissive cells (MORAHAN et al. 1989; WU et al. 1990). The block is prior to synthesis of HSV DNA and the DNA disappears by 24 h after infection (LEARY et al. 1985). However, there is some HSV gene expression in resident peritoneal MOs. Thus, by immunofluorescence, the IE ICP4 protein can be detected in a small proportion of the MOs (SIT et al. 1988; MORAHAN, unpublished). The block in resident peritoneal MOs appears to occur in expression of the HSV early genes and may involve post-transcriptional control in regulating HSV-1 early gene expression in the context of the MOs. Using nuclear run-on assays, transcription of all kinetic classes of HSV-1 genes (IE, early, and late) was efficiently initiated in the nonpermissive resident peritoneal MO from CD-1 mice. However, only steady state levels of mRNAs specifying HSV-1 IE proteins were detected, while expression of some or all of the early genes examined was minimal, and mRNA for the true late gene product, gC, was not detected (MORAHAN et al. 1989).

Dysfunction of at least one of the IE gene products also appears to play a role in the nonpermissiveness of resident peritoneal MOs. Using a transient transfection expression assay, we have investigated both the transactivating activity of HSV-1 IE genes encoding ICPO and ICP4 and the associated promoter activity of the early genes encoding the thymidine kinase and DNA polymerase enzymes. In the nonpermissive CD-1 cells, the IE gene product ICPO appeared to function inefficiently in transactivating the early gene promoters in comparison with ICP4. The SCHAFFER group has elegantly demonstrated that, although

ICPO is not essential for virus replication in cells culture, it is important to modulate viral gene expression (CAI and SCHAFFER 1989). A defect in the upregulatory function in gene expression by ICPO in the context of the resident peritoneal MOs may play a critical role in restricting virus replication. Further examination of the interaction among ICPO, *cis*-acting signals, and cellular transcriptional factors in nonpermissive mouse resident peritoneal MOs will help elucidate the molecular basis of the restriction of HSV-1 replication in these cells.

4.1.2 Block in HSV-1 Replication in Human Monocytes and MOs

The general conclusion is that the intrinsically resistant monocyte may play a critical role in limiting HSV-1 infection in humans, but it must be realized that this type of correlative in vitro data can only be cautiously extended to in vivo resistance. Primary cultures of human monocytes can take up and degrade HSV (VAN STRIJP et al. 1990) and are relatively nonpermissive to HSV replication (reviewed in MORAHAN et al. 1985); however, where the replication block occurs has received little investigation. ALBERS et al. (1989) have observed that no viral DNA was associated with the nuclear fraction of freshly cultured monocytes, indicating that the block in human monocytes is even earlier in replication than the block in mouse resident peritoneal MOs. Human alveolar MOs were shown to be extremely resistant to HSV-1 infection. The same population of cells supported productive infection of cytomegalovirus, suggesting that the mechanisms of MO intrinsic resistance may vary between these two herpesviruses (DREW et al. 1979).

Similar to the studies with mice (JOHNSON 1964), monocytes obtained from human neonates may be more susceptible to HSV infection than monocytes obtained from adults (TROFATTER et al. 1979), but this has not been the experience of all investigators (PLAEGER-MARSHALL et al. 1989).

4.1.3 Role of Endogenous or Early HSV-Induced IFN or Other Cytokines in Intrinsic Resistance

As is clear from the preceding sections, MOs a priori seem to have prominent intrinsic and extrinsic antiviral potentials. Another prominent antiviral system is provided by the IFNs. As MOGENSEN and VIRELIZIER (1987) have stated, it is tempting to speculate that MOs and IFN "may act in concert to constitute an early, interrelated and interdependent defense network." This can be envisaged to occur in two ways. First, MOs, or other cells could be continuously stimulated with IFN inducers (e.g., microbial normal flora), so that MOs in vivo are in a constant IFN-induced antiviral state. Although IFN is often no detected in naive animals and cells, sometimes the IFN-inducible enzymes, such as 2′,5′. A synthetase can be detected in MOs, suggesting that low levels of IFN are present. The second way an IFN mechanism could operate is through very early induction of IFN when MOs are infected with HSV, therefore resulting in rapid induction of the IFN-induced antiviral state.

With some viral infections (e.g., vesicular stomatitis, and encephalomyocarditis viruses) MOs do exhibit an endogenous, IFN-mediated, intrinsic, antiviral resistance. Treatment with antibodies to IFN in vivo decreases both the intrinsic resistance of the MO and the levels of 2′, 5′. A synthetase in peritoneal MOs from normal mice (BELARDELLI et al. 1984). This resistance is probably·produced by normal flora microbial IFN inducers in vivo. A role for endogenous IFN in the intrinsic resistance of resident peritoneal MOs to HSV appears unlikely. Treatment of CD-1 mice anti-IFN antibody did not reverse nonpermissiveness, indicating that endogenous IFN was unlikely to contribute to the intrinsic resistance of CD-1 resident peritoneal MOs to HSV-1 infection (SIT et al. 1988). Early production of IFN by HSV-infected peritoneal MOs, however, may play some role in intrinsic resistance. Treatment with antibodies to IFN, both in vivo and in vitro, slightly decreased the intrinsic resistance of resident peritoneal MOs to HSV, suggesting that virus-induced IFN may play a role (SIT et al. 1988). Early production of IFN by HSV-infected peritoneal MOs has been associated with the X-linked resistance of C57B1/6 mice and the intrinsic resistance of their resident peritoneal MOs to HSV-2 infection (MOGENSEN 1979). HSV-induced IFN or endogenous IFN also activated resistant C57Bl/6 peritoneal MOs for a respiratory burst that was increased compared to the burst of cells from susceptible Balb/c mice and produced an enhanced antiviral state in C57Bl/6 cells as compared with Balb/c mouse cells (ELLERMAN-ERIKSEN et al. 1989). What role an enhanced respiratory burst may play in intrinsic antiviral resistance remains to be determined.

The role that endogenous or early HSV-induced IFN may play in the intrinsic resistance of freshly cultured human monocytes to HSV has also been investigated. Freshly isolated human monocytes are very resistant to HSV infection, and this is correlated with a production of high levels of IFN-α/β that does not require virus replication (ALBERS et al. 1989). ALBERS et al. (1989), however, have concluded that IFN induced by HSV is not of major importance, since HSV

Table 1. Possible roles of IFN in MO-HSV interaction

Antiviral effects
- Induction of (2′–5′) oligoadenylate synthetase, which catalyzes the synthesis of pppA(2′p5′A)n, resulting in activation of RNase L and subsequent degradation of mRNA by RNase L
- Induction of a protein kinase, which phosphorylates eIF-2, resulting in inhibited protein synthesis
- Induction of a phosphodiesterase that removes the sequence CCA from the 3′ end of tRNA molecule, resulting in the inhibition of polypeptide elongation

Cell regulatory effects
- Inhibition of cell proliferation
- Effects on cell differentiation
- Effects on structural cell components
- Effects on immune functions
  - Increase the synthesis and membrane expression of a series of molecules that have a critical role in opsonization, adherence, and antigen presentation
  - Enhance/induce cytokine production
  - Increase intracellular killing
  - Increase release of oxygen radicals

replication was inhibited prior to viral DNA entry into the nucleus and prior to the time for virus-induced IFN to be produced (Table 1).

In summary, the current evidence indicates that the role of IFN in intrinsic antiviral resistance varies depending upon both the source of the MO and the infecting virus. In MOs, there is little evidence for an endogenous, IFN-mediated, antiviral state regarding HSV infection, but there is evidence for early HSV-induced IFN in some MOs. Additional investigation is warranted to establish whether MOs exhibit any common mechanism of intrinsic antiviral activity against different viruses or whether the mechanism varies depending upon the particular virus and MO differentiation or activation state. This information is critical for develop-ing optimum antiviral chemotherapy and immunotherapy that involves manipulation of MOs.

4.2 Modulation by MO Activation or Differentiation and Tissue Source

Mo activation, maturation, or differentiation has profound effects on MO-mediated intrinsic and extrinsic resistance to HSV. An increase occurs in MO-mediated extrinsic resistance that is associated with MO activation/differentiation. Thus, inflammatory or lymphokine-activated MOs could decrease virus burden in surrounding permissive host cells (Sect. 3). in regard to MO intrinsic antiviral resistance, the emerging conclusion is that increased HSV gene expression occurs in activated or differentiated MOs (MORAHAN et al. 1989). Nevertheless, HSV replication is severely limited in all MOs, and thus activated/differentiated MOs can still serve to limit HSV infection in vivo. Furthermore, activated/differentiated MOs that are infected with HSV often show a severe cytopathic effect. This destruction of virally infected activated MOs could play a role in host resistance. Activation is generally accompanied by a large increase in the number of MOs present. These MOs could internalize more virus than could a smaller number of resident MOs. Subsequent MO lysis or "suicide", before production of infectious virus, could decrease the virus infection and remove MOs that have some permissiveness for HSV infection, leaving a smaller infection burden and completely resistant resident MOs.

The terms MO differentiation, maturation, and activation are being used interchangeably, since these processes may occur simultaneously and a clear distinction is difficult to establish. Differentiation can be defined as *permanently altered* cellular gene expression and MO activation as a process that leads to *reversible changes* in phenotype and function. The term maturation is used in various ways; in the context of MO, the term is often used to describe the changes that take place when MO are cultured (aged) in vitro, or the process whereby immature MO lose their potential for cell division. There is also much heterogeneity among MOs depending on their tissue localization. Furthermore, it is increasingly evident that tissue MO populations may be maintained independent of the circulating monocyte population and may have a different embryonic origin (MORAHAN et al. 1988). Elucidating the origins for the diversity among MOs is of obvious importance in establishing the exact roles that certain MOs play in natural, immunomodulator-enhanced, or specific immune responses to HSV infection.

4.2.1 Studies in Experimental Animals

Heterogeneous intrinsic antiviral resistance occurs within the same population of MOs (PLAEGER-MARSHALL et al. 1982; SIT et al. 1988). For example, about 15% of resident peritoneal MOs of B6C3F1 mice expressed the HSV-1 IE antigen ICP4 and synthesized viral mRNAs, while the remaining resident peritoneal MOs were completely nonpermissive (MORAHAN et al. 1989). Yet to be determined are the differences between the semipermissive and completely resistant MOs.

A second level of heterogeneity in intrinsic resistance is the tissue location of the MOs. A study of HSV-1 infection of various rabbit MOs illustrates this point (PLAGER-MARSHALL et al. 1982). The most resistant cell was the monocyte, which neither adsorbed nor replicated HSV. Resident alveolar MOs adsorbed but did not replicate HSV yet were killed. Paraffin oil-elicited-peritoneal MOs were semipermissive, with 40% of the cells expressing viral antigen and a small precentage producing infectious virus. Similar heterogeneity has sometimes, but not always, been noted with mouse resident MOs from various organs (HENDRZAK et al. 1991).

A third level of heterogeneity, accompanied by increased HSV gene expression, occurs when MOs are activated or differentiated by various treatments in vivo or in vitro. Culturing (aging) resident peritoneal MOs in vitro for several days leads to increased permissiveness for HSV replication (MOGENSEN 1979). Treatment of mice with immunomodulators (killed *C. parvum* vaccine, avirulent *S. typhimurium* vaccine) or sterile inflammatory agents (thioglycollate broth) produces activated MOs that show increased permissiveness by increased expression of the IE ICP4 protein and a cytopathic effect (MORAHAN et al. 1985). In addition, an increased production of infectious virus occurred in thioglycollate-activated MOs, although these cells were still not fully permissive (SIT et al. 1988). The intrinsic resistance of MOs also changes when they are activated in vitro. When splenic MOs were activated by treatment with lipopolysaccharide for 2 weeks in vitro, lipid A, or tumor necrosis factor (TNF), there was an increase in HSV yield and a cytopathic effect (DOMKE-OPITZ and KIRCHNER 1990). Interestingly, treatment with interleukin-1 (IL-1) did not enhance permissiveness.

The intrinsic resistance to HSV-1 infection of the differentiated bone marrow-derived MOs (BMD MOs) has received considerable attention. Like activated MOs, BMD MOs, differentiated in the presence of granulocyte/macrophage colony-stimulating factor (GM-CSF), generally show increased permissiveness for HSV as compared with resident peritoneal MOs (LEARY et al. 1985). HVS-1 DNA increased and a cytopathic effect was apparent, although infectious virus was not detected. SARMIENTO (1988) has reported that HSV-1 macromolecular synthesis was inhibited at two levels in GM-CSF-differentiated BMD MOs. This occurred before the onset of IE protein synthesis in the majority of cells and between the onset of DNA synthesis and late protein synthesis in a minority. KRUSE et al. (1989) have confirmed that BMD MOs differentiated with GM-CSF were highly susceptible to lytic infection with HSV but that BMD MOs differentiated with macrophage colony-stimulating factor (M-CSF) were more resistant. This was correlated with autocrine production of IFN by bone marrow cells cultured with M-CSF.

The cytopathic effect observed in many activated MO populations in response to HSV infection, without necessarily the production of infectious virus, is intriguing. It requires infection with replication-competent virus and does not appear to involve cell lysis caused TNF (SIT et al. 1988), although the involvement of other cellular factors or viral gene products remains to be determined.

4.2.2 Studies in Human Monocytes and Monocyte-Derived MOs

Increased permissivity for HSV infection has also been noted with MOs that have differentiated from human monocytes, i.e., monocytes that have been cultured (aged) in vitro for several days and have differentiated into MOs. Loss of capacity for production of virus-induced IFN has been suggested as a mechanism, but IFN-independent mechanisms are also viewed as likely (ALBERS et al. 1989).

4.2.3 Studies in Monocyte-MO Cell Lines

The heterogeneity of MOs and the difficulty in obtaining sufficient cells from primary sources led to the development of monocyte-MO-like cell lines (HARRIS and RALPH 1985). Both mouse and human MO-like cell lines have been used to investigate the effect of differentiation/activation on the intrinsic interaction between MOs and HSV. In general, it appears that differentiation increases the permissiveness of MO cell lines for HSV infection. Fox example, undifferentiated MO-like U937 cells are very resistant to HSV-1 infection; however (TENNEY and MORAHAN 1991; KEMP et al. 1990), after exposure to phorbol myristate acetate (PMA) or vitamin D_3 HSV-1 viral protein synthesis and infectious virus production increased significantly (TENNEY and MORAHAN 1991). The mechanism of increased permissiveness appears associated with post-transcriptional regulation of IE gene expression (KEMP et al. 1990), but transcriptional regulation of IE genes has also been suggested (TENNEY and MORAHAN 1991). HSV may also persist in undifferentiated U937 cells until the block in replication is released by subsequent differentiation (TENNEY and MORAHAN 1991). SALO and ORTEGA (1986) have concluded that IFN is not involved in resistance of undifferentiated MO-like cell lines because addition of anti-IFN serum did not abrogate resistance.

In contrast to U937 cells, HL-60 cells differentiate into MO-like cells when treated with PMA and into PMNs when treated with dimethyl sulfoxide (DMSO) (HUBERMAN and CALLAHAM 1979). HSV-1 replicated equally well in undifferentiated or PMA-differentiated HL-60 cells, but in DMSO-differentiated cells abortive infection occurred, with the block apparently at the level of early gene expression (PIENTONG et al. 1989). These data suggest that permissiveness for HSV may vary between the MO and granulocyte cellular environments.

A persistent infection of HSV-1 has been established in an Ia antigen-bearing mouse MO cell line, SL-1 (TANG et al. 1988). Persistent infection could not be established in a non-Ia antigen-bearing SL-I cell line, suggesting a requirement for activation-related cellular factors, at least in regard to Ia antigen expression. Heterogeneity of permissiveness was apparent; persistent infection occurred in a minority of cells with the majority uninfected. Viral DNA structure was altered in the persistently infected cells as revealed by restriction enzyme analysis.

The possibility that MOs may harbor HSV and transmit the virus when they are further activated or differentiated is of obvious importance in persistent or latent HSV infection in vivo and merits additional investigation. Current development of molecular techniques, such as subtractive hybridization, will allow better definition of the unique factors that appear with differentiation and their role in HSV gene expression.

4.2.4 Role of IFN in Regulating MO Differentiation and HSV Infection

There are three possible sources whereby IFN may be produced and affect MOs: (1) endogenous IFN produced in vivo, which leads to a MO IFN—induced antiviral state; (2) early IFN induced by HSV infection in MOs; and (3) exogenous IFN produced by other cells, e.g., IFN-γ produced by lymphocytes, that then affect MOs. Whatever the source and type of IFN, the consensus is that IFN causes pleiotropic effects on MOs (Table 1).

In Sect. 4.1, the possible role of endogenous or early virus-induced IFN in mediating intrinsic antiviral effects in MOs was discussed. In this section, we will briefly review the effects of exogenous or virus-induced IFN on MO differentiation and antiviral activities. Several mechanisms have been described, varying with the particular type of MO and IFN. For example, exogenous IFN-γ and IFN-α/β appear to inhibit HSV-1 replication in mouse splenic MOs in different ways. IFN-α/β treatment of HSV-1-infected mouse spleen MOs cultured in the presence of CSF inhibited the genes encoding IE and early gene products at the transcriptional level (STRAUB et al. 1986), while IFN-γ, inhibited the same genes at the translational level (KLOTZBUCHER et al. 1990). At this point, it is unclear whether all MOs use a common IFN antiviral mechanism or whether this varies depending upon the particular MO and virus interaction.

It remains to be established whether IFN-induced antiviral effects in MOs are mediated by the two major IFN antiviral systems (Table 1) or by regulatory activities of IFN on cell proliferation, structural cell components, differentiation, and immune functions (JOKLIK 1990). Moreover, there are questions as to whether the known IFN antiviral mechanisms also serve some other function, such as during cell differentiation. The role of IFN in regulating MO-HSV interaction has been extensively studied in the differentiation of human monocytes to monocyte-derived MOs, which is associated with a decrease in intrinsic resistance to HSV (discussed previously). It has been clearly shown that either exogenous IFN treatment or abortive infection with high multiplicities of HSV which induces IFN can inhibit the differentiation of monocytes into MOs and maintain intrinsic resistance to HSV (LINNAVUORI and HOVI 1987). IFN may also inhibit, under certain conditions, the proliferation and differentiation of other MOs, such as BMD MOs in response to CSF (MOORE et al. 1984). It is also clear that monocyte-derived MOs, which appear to be defective in IFN induction by HSV and are susceptible to HSV replication and cytopathic effects (ALBERS et al. 1989), can still be protected by treatment with exogenous IFN (LAZDINS et al. 1990). It is obvious that more investigation is required in order to delineate the molecular mechanisms involved in IFN's effect on MO proliferation and differentiation and on HSV replication.

5 Other Nonspecific Mechanisms that Provide Resistance to HSV Infection

Nonspecific resistance other than that provided by MOs and IFN include NK cells, PMNs, and Langerhans' cells.

5.1 Role of PMNs

Early studies indicated that HSV-infected cells produced factors chemotactic for PMNs (SNYDERMAN et al. 1972) and that PMNs could inhibit bovine HSV-1 replication and cause antibody-dependent cellular cytotoxicity (ADCC) (ROUSE et al. 1980). More recently, human PMNs were demonstrated to exhibit both intrinsic and extrinsic resistance to HSV infection (VAN STRIJP et al. 1989a, b).

Regarding intrinsic resistance, as many as 10000 viral particles could be associated with one PMN through Fc-mediated phagocytosis. The resistance of PMNs, unlike that of MOs, required virus-specific antibody and was enhanced by complement (VAN STRIJP et al. 1989a, b). Antibody-independent complement-dependent phagocytosis also occured if the PMNs were previously primed with TNF (VAN STRIJP and colleagues (1991) have proposed that a novel nonspecific resistance to HSV is provided by PMNs, together with TNF or GM-CSF, viral-specific antibody, and complement. The whole process was assumed to be regulated by monocytes that produce TNF in response to HSV infection. Interestingly, TNF production could not be detected in mouse peritoneal MOs infected with HSV (SIT et al. 1988). This may indicate a difference in species or MO differentiation.

A number of studies have addressed the mechanisms of intrinsic antiviral resistance of PMNs to HSV. The interaction of PMNs with HSV-antibody complexes caused a gradual chemiluminescence response (BINGHAM et al. 1985), but oxygen radicals appear unnecessary for antiviral activity (VAN STRIJP et al. 1990). Another oxygen-independent antiviral mechanism described recently in PMNs is the cationic defensin peptides, which appear to directly inactivate HSV and other viruses (LEHRER et al. 1985). Mediators produced by bovine PMNs, possibly similar to defensins, have also been described (ROUSE et al. 1980).

PMNs can also exhibit extrinsic antiviral resistance via ADCC. Peripheral blood PMNs efficiently bound to HSV-infected fibroblasts in the presence of antibody or complement and mediated cytotoxic effects (VAN STRIJP et al. 1989a; ASHKENAZI and KOHL 1990).

The exact role of PMN-mediated antiviral mechanisms in vivo remains to be established. We have observed a strong, early, peritoneal PMN inflammatory response associated with intraperitoneal HSV-1 infection in mice but not with HSV-2 infection (HENDRZAK, unpublished observations). Infiltration of PMNs and phagocytosis of free virions by PMNs occur in experimental keratitis and in human HSV-2 lesions, implying a role for these phenomena in vivo (BODDINGIUS et al. 1987). The protective or possible immunopathological roles of PMNs in inflammatory responses to HSV infection in sites such as the cornea remain to be elucidated (see the chapter by DOYMAS and ROUSE).

5.2 Role of NK Cells

The role of NK cells in HSV infections remains uncertain and currently receives little investigation. The few studies of intrinsic resistance of NK cells to HSV infection indicate that NK cells are quite nonpermissive, similar to MOs (HAYWARD et al. 1989). Many reports have indicated that NK cells can exhibit extrinsic antiviral resistance by lysing HSV-infected cells or inhibiting HSV replication in infected target cells (see review WELSH and VARGAS-CORTES 1991). The mechanisms involved at both the effector cell and the virus-infected target cell level have been explored. In the overnight, in vitro, NK cytotoxicity assay, NK cells may be activated by IFN production by another accessory cell, perhaps a dendritic cell (FITZGERALD-BOCARSLY et al. 1988). At the target cell level, expression of some of the HSV early and possibly late genes is required to produce sensitivity of the infected cells to NK-mediated lysis (WELSH and CORTES 1991). The use of both viral mutants and cells transfected with vectors expressing viral gene products will be helpful in elucidating further the viral gene products required for sensitization of the infected target cells or activation of the effector cells.

Whether NK cells play a role in resistance to human HSV infections remains uncertain. It is clear that HSV infections are more common and severe in several immunocompromised states such as found in neonates, genetically immunodeficient states, and associated with immunocompromising diseases such as Hodgkin's disease and AIDS (WELSH and VARGAS-CORTES 1991). NK cells are clearly depressed in these clinical situations, although other specific and nonspecific effectors may also be depressed. Neonatal NK cells have impaired cytotoxicity for HSV-infected target cells (LEIBSON et al. 1986). Also, NK cells from pregnant women have an attenuated response against HSV-infected target cells (GONIK et al. 1990).

In mice, the emerging consensus is that NK cells do not play a major role in the natural resistance of normal mice to acute infection with HSV. Early reports indicated that resistance to HSV infection was reduced by prolonged treatment of mice with ^{89}Sr (LOPEZ et al. 1980) or with high levels of antibody to- asialo-GM1 (HABU et al. 1984). However, other studies have shown that resistance to HSV infection was normal in NK-deficient beige mice (BUKOWSKI and WELSH 1986). Resistance was also normal in mice that were treated with anti-asialo-GM1 or monoclonal antibody to the NK 1.1 antigen in concentrations sufficient to deplete NK cells but not the early IFN response (BUKOWSKI and WELSH 1986).

Adoptive cell transfer experiments have also indicated that NK cells are not required in the natural resistance of normal mice to HSV infections. Transfer of adult leukocytes to suckling mice provided antiviral protection, but removal of the NK cells did not abrogate the enhanced resistance (BUKOWSKI and WELSH 1986). However, adoptive transfer of cell into cyclophosphamide-treated mice did appear to require the presence of NK cells for the protective antiviral effects (RAGER-ZISMAN et al. 1987).

The various observations may be interpreted to indicate the normal redundancy of natural immune responses, which is provided by the interaction of MOs, early induction of IFN, and NK cells in the normal resistance of the mouse to HSV infections. In immunocompetent hosts infected with HSV, the degree of early IFN-induced resistance by MOs (and/or other cells) is sufficient so that NK cell are not required. However, in certain types of immunosuppression (e.g., cyclophosphamide-induced) there may be impairment of the early IFN response may be such that NK cells may take on a greater role. Systematic use of selective depletions of MOs, NK cells, and IFN, in conjunction with careful molecular pathogenesis studies, may help elucidate the complex interactions of these nonspecific effectors.

5.3 Role of Langerhans' Cells

Langerhans' cells (LCs) are dendritic cells of bone marrow origin, possibly related to the mononuclear phagocyte lineage, which are located in epithelial tissues and function as antigen-presenting cells. Their role has been investigated in HSV infections of the cornea and skin using kinetic correlative, cell depletion, and adoptive cell transfer methodologies. Depletion of LCs from the skin, by treatment with 10% aqueous saline and abrasion, led to enhancement of virulence following footpad infection of mice with a nonpathogenic HSV strain (SPRECHER and BECKER 1987). Depletion of epidermal LCs by in vivo exposure of mice to UV light impaired the resistance of mice to intracutaneous infection with HSV (OTANI and MORI 1987). Exposure of mice to UV light also led to the depressed capacity of epidermal LCs to present HSV-1 antigen in an in vitro HSV antibody induction system (HOWIE et al. 1986).

Adoptive cell transfer studies have also shown a role for LCs in resistance to cutaneous HSV infection. Intracutaneous transfer of spleen cells, stimulated by culture with epidermal LCs and UV-inactivated HSV, reduced the virus titer in the skin and the formation of zosteri form skin lesions in nude mice infected intracutaneously with HSV (YASUMOTO et al. 1986). Exposure of mice to UV light, prior to isolation of the epidermal LCs reduced the ability of the cells to stimulate spleen cells and to transfer resistance to nude mice (OTANI and MORI 1987). UV exposure prior to cell transfer, treatment of mice with *cis*-uroconic acid (which appears to be the mediator induced by UV exposure), or transfer of LCs from neonatal mice generate suppression of delayed-type hypersensitivity to HSV (HOWIE et al. 1987). These results indicate that LC-mediated resistance is age-dependent, as is MO-mediated resistance.

The host response to corneal infection with HSV was also shown to involve LCs. Briefly, histopathological studies have demonstrated the accumulation of LCs at the sites of corneal infection (LEWKOWICZ-MOSS et al. 1985). The interrelationships of the LC response with PMNs and mononuclear inflammatory responses, the modulatory role of UV light, and the protective vs immunopathogenic role of LCs in corneal infection are under investigation in a number of laboratories.

5.4 Cytokines and Other Natural Resistance Mechanisms

It is evident that the antiviral effects of MOs, PMNs, and NK and other cells may be modulated by cytokines produced by these or other cells during HSV infection. The IFN family of cytokines has received the most attention, but there are a few reports involving others. In vitro, NK cell-mediated lysis of HSV-infected target cells is apparently increased by activation of the cells with IL-2 (LEIBSON et al. 1986). PMN destruction of HSV was enhanced by activation of PMNs with TNF (VAN STRIJP et al. 1991).

The role of cytokines in vivo has received less attention. Again, the IFN family of cytokines appears to be important in natural resistance, although whether the effect is direct or indirect through activation of other cells remains unclear. Transfer of neonatal leukocytes together with IL-2 produced protection against HSV infection of neonatal mice (KOHL 1990). Depletion of either IFN-α/β or -γ by treatment of mice with antibodies to IFN caused a decrease in resistance (STANTON et al. 1987; GRESSER et al. 1976). Furthermore, treatment of mice with exogenous IFN -α, -β, or -γ provided strong protection against HSV infection.

Repeated prophylactic and therapeutic treatment of adult mice with exogenous cytokines such as IL-1, CSF-1, and GM–CSF have failed to enhance resistance to HSV infection (MORAHAN and PINTO, unpublished observations). This may merely reflect the normal redundancy of natural effector systems in immunocompetent adult mice or the need for several interacting cytokines and effector cells for enhanced protection. It is clear that additional investigations, such as depletion of cytokines with monoclonal antibody treatment or administration of cytokines, will be necessary to elucidate their role in resistance to HSV infections. Some cytokines are known to be produced endogenously and therefore could lead to a cytokine-induced antiviral state. It remains to be established whether any of these cytokines are induced sufficiently early after HSV infection of MOs to provide antiviral resistance.

A number of other nonspecific resistance mechanisms were demonstrated to be important in certain viral infections (MORAHAN and MURASKO 1989). These include the effects of nutritional status, sex, pregnancy, normal microbial flora, body temperature, stress, and other neurohormones. Except for the well documented increase in resistance to HSV that is associated with age, in both experimental animal systems and in humans little attention has been paid to the role of these various factors in host resistance to HSV.

6 Suppressive Effects of HSV Infection on Functions of Nonspecific Immune Effector Cells

It seems that virus infection can alter the biological or "luxury" functions of host cells, leading to either enhancement or suppression of normal functions in the absence of any observable morphological damage or "cellular injury" (OLDSTONE 1989). The suppressive effects of two herpesviruses, cytomegalovirus and Epstein-Barr virus, on the functions of immune cells were clearly demonstrated (RINALDO 1990). With HSV infection, numerous MO functions can be changed, including inhibition of chemotaxis (KLEINERMAN et al. 1974) and decreased ADCC and phagocytosis (PLAEGER-MARSHALL et al. 1983). Infection of mouse BMD MOs with HSV-1 induced expression of Ia antigen on infected cells but with no changes in phagocytosis or FC receptor expression (HOWIE et al. 1986). A pronounced inhibitory effect of HSV-1 infection on IFN-γ activation of mouse peritoneal MO tumoricidal activity which required abortive virus infection was proposed to be mediated by virus-induced IFN-α (KOFF et al. 1987). Using northern blot hybridization and the polymerase chain reaction (PCR), we recently observed that mRNA encoding mouse IL-1β was sharply reduced as early as 6 h after HSV-1 infection of *C. parvum*-or avirulent *S. typhimurium*-activated pertioneal MOs (WU et al. 1991). The mechanism of altered cytokine gene expression caused by HSV-1 infection of activated peritoneal MOs is currently under investigation.

Recently, it has been reported that HSV-1 can reactivate a latent infection of another virus in vitro (MOSCA et al. 1987). Using a superinfection assay, HSV-1 gene expression was shown to reactivate transcription of latent HIV infection in cell lines permanantly transfected with the HIV long terminal repeat (LTR) (MOSCA et al. 1987). Using deletion mutants within the HIV LTR, the targets for HSV stimulation were shown to be different from the *tat*-responsive area and mapped near the SPI binding sites (OSTROVE et al. 1987). If similar results are found with latently infected MOs or neurons, there is an obvious implication for the reactivation of HIV in latently infected MOs by HSV-1 infection.

During HSV infection in vivo, both inhibition of PMN chemotactic activity and depressed oxidative metabolism in response to zymosan stimulation but normal bactericidal action have been reported in infected animals and humans (OHMANN and BABIUK 1985; ABRAMSON and MILLS 1988). NK cell activity in calves was depressed after bovine HSV-1 infection (OHMANN and BABIUK 1985). Severely suppressed lymphocyte blastogenesis and cytokine production may occur during acute HSV-1 infections of humans (SHERIDAN et al. 1987; RINALDO 1990). It must also be noted that HSV infection can lead to recruitment of various immune cells with immunopathological results, such as demyelination or corneal damage (TOWNSEND and COLLINS 1986). With the exact roles of nonspecific effectors, such as MOs, PMNs, LCs, and cytokines, may be in such immunopathology remains to be elucidated.

In conclusion, while the informations is still rudimentary, HSV infection can induce functional impairment in nonspecific immune effector cells in the absence of any morphological damage. The observed immune suppression by HSV infection may not necessarily correlate with the intrinsic resistance of these cells to virus infection. It remains to be determined whether the suppression is caused by viral gene products themselves or is a result of the altered physiological state of these cells by virus infection.

7 Conclusions and Future Perspectives

A variety of nonspecific effector cells and cytokine systems—MOs, NK cells, PMNs, and IFNs—have clearly been demonstrated to exhibit intrinsic and/or extrinsic antiviral activity in vitro against HSV and HSV-infected cells. An understanding of the molecular mechanisms involved in these antiviral activities in vitro is beginning to emerge. Correlative kinetic studies and selective cell depletion and adoptive transfer studies have established antiviral activity in vivo in HSV infections in several experimental animal models.

There is now a well founded concept for complex interactions of the natural immunity systems, both among themselves and with specific cellular and humoral anti-HSV responses. The dynamic, changing, temporal processes associated with different target organs during pathogenesis are becoming apparent. Elucidation of these complexities will be aided by the application of modern molecular approaches to the study of the natural immune responses. The use of techniques such as: (a) in situ reverse transcription PCR, to detect viral and cytokine mRNA expression, (b) viral mutants differing in selected genes, (c) transgenic mice expressing selective cytokines or viral genes, or (d) transgenic mice bearing genes to selectively inactivate cells or cytokines should prove useful. More investigations of combined depletion of two or more effectors, such as NK cells and IFN, will help elucidate the unique interactions that may be required for optimal natural host resistance.

How important these natural immune systems are in the HSV infections of humans remains to be resolved. Nevertheless, considerable correlative data, especially in immunocompromised clinical situations, provide evidence for their potential significance. The potent natural resistance amplification systems are likely to be useful as major weapons in host defense against HSV and other viral infections in the emerging development of immunotherapy for infectious diseases.

Acknowledgment. This work was supported by Public Health Service grants AI 25751 and AI 24004 from the National Institute of Allergy and Infectious Diseases.

References

Abramson JS, Mills EL (1988) Depression of neutrophil function induction induced by viruses and its role in secondary microbial infections. Rev Infect Dis 10: 326–344

Albers I, Kirchner, Domke-Opitz I (1989) Resistance of human blood monocytes to infection with herpes simplex virus. Virology 169: 466–469

Ashkenazi M, Kohl S (1990) Reduced antibody-dependent cellular cytotoxicity to herpes simplex virus-infected cells of salivary polymorphonuclear leukocytes and inhibition of peripheral blood polymorphonuclear leukocyte cytotoxicity by saliva. J Immunol 144: 4781–4787

Belardelli F, Vignaux F, Proietti E, Gresser I (1984) Injection of mice with antibody to interferon renders peritoneal macrophages permissive for vesicular stomatitis virus and encephalomyelitis virus. Proc Natl Acad Sci USA 81: 602–606

Ben-Hur T, Rosen-Wolff A, Lamade W, Darai G, Becker Y (1988) HSV-1 DNA sequence determining intraperitoneal pathogenicity in mice is required for transcription of viral immediate-early genes in macrophages. Virology 163: 397–404

Bingham EL, Fenger TW, Sugar A, Smith JW (1985) Dependence on antibody for production of chemiluminescence in polymorphonuclear leukocuytes by herpes simplex virus. Invest Opthalmol Vis Sci 26: 1236–1243

Boddingius J, Dijkman H, Hendriksen E, Schift R, Stolz E (1987) HSV-2 replication sites, monocyte and lymphocytic cell infection and virion phagocytosis by neutrophils, in vesicular lesions on penile skin. J Cutan Pathol 14: 165–175

Bukowski JF, Welsh RM (1986) The role of natural killer cells and interferon in resistance to acute infection of mice with herpes simplex virus type 1. J Immunol 136: 3481–3485

Cai W, Schaffer PA (1989) Herpes simplex virus type 1 ICPO plays a critical role in the de novo synthesis of infections virus following transfection of viral DNA. J Virol 63: 4579–4589

Connell, EV, Cerruti RL, Trown PW (1985) Synergistic activity of combinations of recombinant human alpha interferon and acyclovir, administered concomittantly and in sequence, against a lethal herpes simplex virus type 2 infection in mice. Antimicrob Agents Chemother 28: 1–4

Domke-Opitz I, Kirchner H (1990) Stimulation of macrophages by endotoxin results in the reactivation of a persistent herpes simplex virus infection. Scand J Immunol 32: 69–75

Drew WL, Mintz L, Hoo R, Finley TN (1979) Growth of herpes simplex virus and cytomegalovirus in cultured human alveolar macrophages. Am Rev Respir Dis 119: 287–291

Ellermann-Eriksen S, Sommerlund M, Mogensen SC (1989) Differential sensitivity of macrophages from herpes simplex virus-resistant and-susceptible mice to respiratory burst priming by interferon-α/β. J Gen Virol 70: 2139–2147

Fitzgerald-Bocarsly P, Feldman M, Mendelsohn M, Carl S, Lopez C (1988) Human mononuclear cells which produce interferon-alpha during NK (HSV-FS) assay are HLA-DR positive cells distinct from cytplytic natural killer effectors. J Leukocyte Biol 43: 323–334

Gonik B, Loftin KC, Tan NS, Crump J (1990) Immune modulation of natural killer cell cytotoxicity against herpes infected target cells in pregnancy. Am J Reprod Immunol 24: 95–98

Greeser I, Tovey MG, Maury C, Bandu MT (1976) Role of interferon in the pathogenesis of virus diseases in mice as demonstrated by the use of anti-interferon serum. II. Studies with herpes simplex, Moloney sarcoma, vesicular stomatitis, Newcastle disease, and influenza viruses. J Exp Med 144: 1316–1323

Habu S, Akamatsu K, Tamaoki N, Okumura K (1984) In vivo significance of NK cells on resistance against virus (HSV-1) infections in mice. J Immunol 133: 2743–2747

Harris P, Ralph P (1985) Human leukemia models of myelo-monocytic development: A review of the HL-60 and U937 cell lines. J Leukocyte Biol 37: 407–422

Hayashi K, Kurata T, Morishima T, Nassery T (1980) Analysis of the inhibitory effect of peritoneal macrophages on the spread of herpes simplex virus. Infect Immun 28: 350–358

Hayward A, Laszlo M, Vafai A (1989) Human newborn natural killer cell responses to activation by monoclonal antibodies. J Immunol 142: 1139–1143

Hendrzak JA, Pinto AJ, Morahan PS (1991) Intrinsic resistance to HSV-1 infection in liver Kupffer cells and peritoneal macrophages from normal and immunomodulator-treated mice. Nat Immun Cell Growth Regul (in press)

Howie SEM, Norval M, Maingay JP (1986) Alterations in epidermal handling of HSV-1 antigens in vitro induced by in vivo exposure to UV-B light. Immunology 57: 225–230

Howie SEM, Ross JA, Norval M, Maingay JP (1987) In vivo modulation of antigen presentation generates Ts rather than T_{DH} in HSV-1 infection. Immunology 60: 419–423

Huberman E, Callaham MF (1979) Induction of terminal differentiation in human promonocytic leukemia cells by tumor promoting agents. Proc Natl Acad Sci USA 76: 1293–1297

Izumi KM, Stevens JG (1990) Molecular and biological characterization of a herpes simplex virus type 1 (HSV-1) neuroinvasiveness gene. J Exp Med 172: 487–496

Johnson RT (1964) The pathogenesis of herpes virus encephalitis. II. A cellular basis for the development of resistance with age. J Exp Med 120: 359–374

Joklik WK (1990) Interferons. In: Fields BN, Knipe DM et al. (eds) Virology, 2nd edn. Raven, New York, pp 393–410

Kaplan, AM, J Brown, JM Collins, PS Morahan, MJ Snodgrass (1978) Mechanism of macrophage-mediated tumor cell cytotoxicity. J Immunol 121: 1781–1789

Kemp LM, Estridge JK, Brennan A, Katz DR, Latchman DS (1990) Mononuclear phagocytes and HSV-1 infection: increased permissivity in differentiated U937 cells is mediated by post-transcriptional regulation of viral immediate-early gene expression. J Leukocyte Biol 47: 483–489

Kleinerman ES, Snyderman R, Daniels CA (1974) Depression of human monocyte chemotaxis by herpes simplex virus and influenza viruses. J Immunol 113: 1562–1567

Klotzbucher A, Mittnacht S, Kirchner H, Jacobsen H (1990) Different effects of interferon-γ and interferon-α/β on immediate early gene expression of HSV-1. Virology 197: 487–491

Koff WC, Showalter SD, Seniff DA, Hampar B (1983) Lysis of herpes-infected cells by macrophages activated with free or liposome-encapsulated lymphokine produced by a mouse T cell hybridoma. Infect Immun 42: 1067–1072

Koff WC, Dunegan MA, Chakrabarty MK, Hampar B, Showalter SD (1987) Herpes simplex virus-induced suppression of macrophage-mediated tumoricidal activity in mouse macrophages. Cancer Res 47:.1534–1537

Kohl S (1990) Protection against murine neonatal herpes simplex virus infection by lymphokine-treated human leukocytes. J Immunol 144: 307–312

Kruse A, Kirchner, H, Zawatzky R, Domke-Opitz I (1989) In vitro development of bone-marrow-derived macrophages. Scand J Immunol 30: 731–740

Lazdins J, Alteri E, Cook KW, Burgin C, Gangemi JD (1990) Use of human monocytes in the evaluation of antiviral drugs: quantitation of HSV-1 cytopathic effects. Antiviral Res 13: 175–186

Leary K, Connor JR, Morahan PS (1985) Comparison of herpes simplex virus type 1 DNA replication and virus production in mouse bone marrow derived and resident peritoneal macrophages. J Gen Viral 66: 1123–1129

Lehrer RI, Daher K, Ganz T, Selsted ME (1985) Direct inactivation of viruses by MCP-1 and MCP-2 natural peptide antibiotics from rabbit leukocytes. J Virol 54: 467–472

Leibson PJ, Hunter-Laszlo M, Hayward AR (1986) Inhibition of herpes simplex virus type 1 replication in fibroblast cultures by human blood mononuclear cells. J Virol 57: 976–982

Lewkowicz-Moss SJ, Shimeld C, Easty DL (1985) The response of Langerhans' cells to a localized corneal inoculation of herpes simplex virus (HSV-1) in the mouse. Opthalmic Res 17: 202

Linnavuori K, Hovi (1987) Herpes simplex virus as an inducer of interferon in human monocyte cultures. Antiviral Res 8: 201–208

Lodmell DL, Niwa A, Hayashi K, Notkins AL (1973) Prevention of cell-to-cell spread of herpes simplex virus by leukocytes. J Exp Med 137: 706–720
Lopez C, Ryshke R, Bennet M (1980) Marrow dependent cells depleted by ^{89}Sr mediate genetic resistance to herpes simplex virus type 1 infections in mice. Infect Immun 28: 1028–1032
Mogensen SC (1979) Role of macrophages in natural resistance to virus infection. Microbiol Rev 43: 1–26
Mogensen SC, Virelizier JL (1987) The interferon-macrophage alliance. Interferon 8: 55–80
Moore RN, Larsen HS, Horohov DW, Rouse BT (1984) Endogenous regulation of macrophage proliferative expansion by colony stimulating factor-induced interferon, Science 223: 178–181
Morahan PS, Morse SS (1979) Macrophage-virus interactions. In: Proffitt M(ed) Virus-lymphocyte interactions: implications: for disease. Elsevier, New York, pp 17–35
Morahan PS, Murasko DM (1989) Viral infections. In: Nelson D(ed) Natural immunity. Academic, Sydney, pp 557–586
Morahan PS, Pinto AJ (1991) An historic overview of biologic response modifiers as antivirals. Can J Infect Dis (in press)
Morahan PS, Mama S, Anaraki F, Leary K (1989) Molecular localization of abortive infection of resident peritoneal macrophages by herpes simplex virus type 1. J Virol 63 2300–2307
Morahan PS, Morse SS, McGeorge MB (1980) Macrophage extrinsic antiviral activity during herpes simplex virus infection. J Gen Virol 46: 291–300
Morahan PS, Connor JR, Leary KR (1985) Viruses and the versatile macrophages. Br Med Bull 41: 15–21
Morahan PS, Dempsy WL, Volkman A, Connor J (1986) Antimicrobial activity of various immunomodulators: independence from normal levels of circulating monocytes and natural killer cells. Infect Immun 51: 87–93
Morahan PS, Volkman A, Melnicoff M, Dempsey WL (1988) Macrophage heterogeneity. In: Heppner G, Fulton A (eds) Macrophages and cancer. CRC Press, Boca Raton, pp 1–25
Morse SS, Morahan PS (1981) Activated macrophages mediate interferon-independent inhibition of herpes simplex virus. Cell Immunol 58: 72–84
Mosca JD, Bednarik DP, Raj NBK, Rosen CA, Sodroski JG, Haseltine WA, Pitha PM (1987) Herpes simplex virus type-1 can reactivate transcription of latent human immunodeficiency virus. Nature 325: 67–70
Oldstone MBA (1989) Virus can cause disease in the absence of morphological evidence of cell injury: implication for uncovering new disease in the fulture. J Infect Dis 159: 384–389
Ohmann HB, Babiuk LAS (1985) Viral-bacterial pneumonia in calves: effect of bovine herpesvirus-1 on immunologic functions. J Infect Dis 151: 937–947
Ostrove JM, Leonard J, Weck KE, Rabson AB, Gendelman HE (1987) Activation of the human immunodeficiency virus by herpes simplex virus type 1. J Virol 61: 3726–3732
Otani T, Mori R (1987) The effects of ultraviolet irradiation of the skin on herpes simplex virus infection: alteration in immune function mediated by epidermal cells and the course of infection. Arch Virol 96: 1–15
Pientong C, Weisshart K, Kuhn JE, Knopf C, Braun RW (1989) Replication of herpes simplex virus type 1 in differentiated human promyelocytic HL-60 cells. Virology 170: 468–476
Pinto AJ, Stewart D, van Rooijen N, Morahan PS (1991) Selective depletion of liver and splenic macrophages using liposomes encapsulating the drug dichloromethylene diphosphonate: effects on antimicrobial resistance. J Leukocyte Biol (in press)
Plaeger-Marshall S, Wilson LA, Smith JW (1982) Permissiveness of rabbit monocytes and macrophages for herpes simplex virus type 1. Infect Immun 35: 151–156
Plaeger-Marshall S, Wilson LA, Smith JW (1983) Alteration of rabbit alveolar and peritoneal macrophage function by herpes simplex virus. Infect Immun 41: 1376–1379
Plaeger-Marshall S, Auk BJ, Altenburger KM, Pizer LI, Johnston Jr RB, Stiehm ER (1989) Replication of herpes simplex virus in blood monocytes and placental macrophages from human neonates. Pediatr Res 26: 135-139
Rager-Zisman B, Quan PC, Rosner M, Moller JR, Bloom BR (1987) Role of NK cells in protection of mice against herpes simplex virus-1 infection. J Immunol 138: 884–883
Rinaldo CR Jr (1990) Immune suppression by herpesviruses. Annu Rev Med 41: 331–338
Rosen H, Gordon S (1990) Adoptive transfer of fluorescence-lebelled cells shows that resident peritoneal macrophages are able to migrate into specialized lymphoid organs and inflammatory sites in the mouse. Eur J Immunol 20: 1251–1258

Rouse BT, Babiuk LA, Henson PM (1980) Neutrophils in antiviral immunity: inhibition of virus replication by a mediator produced by bovine neutrophils. J Infect Dis 141: 223–232

Salo RJ, Orgeta AP (1986) Effect of interferon on herpes simplex virus replication in mouse macrophage-like cells lines. Antiviral Res 6: 161–169

Sarmiento M (1988) Intrinsic resistance to viral infection: mouse macrophage restriction of herpes simplex virus replication. J Immunol 141: 2740–2748

Schirmacher P, Worsdorfer M, Lubbe K, Falke D, Thoenes W, Dienes HP (1989) HSV hepatitis in the mouse: a light and electron microscopic study with immunohistology and in situ hybridization. Virchows Arch [B] 56: 351–361

Sheridan JF, Beck M, Smith CC, Aurelian L (1987) Reactivation of herpes simplex virus is associated with production of a low molecular weight factor that inhibits lymphokine activity in vitro. J Immunol 138: 1234–1239

Sit MF, Tenney DJ, Rothstein JL, Morahan PS (1988) Effect of marophage activation on resistance of mouse peritoneal macrophages to infection with herpes simplex virus type 1 and 2. J Gen Virol 69: 1999–2010

Small JA, Bieberich C, Ghotbi Z, Hess J, Scangos J, Clements JE (1989) The visna virus long terminal repeat directs expression of a reporter gene in activated macrophages, lymphocytes, and the central nervous system of transgenic mice. J Virol 63: 1891–1896

Snyderman R, Wohlenberg C, Notkins AL (1972) Inflammation and viral infection: chemotactic activity resulting from the interaction of antiviral antibody and complement with cells infected with herpes simplex virus. J Infect Dis 126: 207–209

Sprecher E, Becker Y (1987) Skin Langerhans' cells play essential role in the defense against HSV-1 infection. Arch Virol 91: 341–349

Stanton GJ, Jordan C, Hart A, Heard, Langford MP, Baron S (1987) Nondetectable levels of interferon gamma is a critical host defence during the first day of herpes simplex virus infection. Microb Pathog is 3: 179–183

Stevens JG, Cook ML (1979) Restriction of herpes simplex virus by macrophages. An analysis of the cell-virus interaction. Am J Pathol 19–38

Straub P, Domke I, Kirchner H, Jacobsen H, Panet A (1986) Synthesis of herpes simplex virus proteins and nucleic acids in interferon-treated macrophages. Virology 150: 411–418

Tang JL, Yamamoto M, Sakuma S, Mori R, Nagayama (1988) Presistent infection with herpes simplex virus type 1 in an Ia antigen-positive mouse macrophage cell line. Microbiol Immunol 32: 363–374

Tenney DJ, Morahan PS (1991) Differentiation of the U937 macrophage cell line removes an early block of HSV-1 infection. Viral Immunol 4: 91–102

Trofatter KF Jr, Daniels CA, Williams RJ Jr, Gall SA (1979) Growth of type 2 herpes simplex virus in newborn and adult mononuclear leukocytes. Intervirology 11: 117–123

Townsend JJ, Collins PK (1986) Peripheral nervous system demyelination with herpes simplex virus. J Neuropathol Exp Neurol 45: 419–425

Van Strijp JAG, Van Kessel KPM, Van Der Tol ME, Verhoef J (1989a) Phagocytosis of herpes simplex virions by human granulocytes and monocytes. Arch Virol 104: 287–298

Van Strijp JAG, Van Kessel KPM, Van Del Tol ME, Verhoef J (1989b) Complement-mediated phagocytosis of herpes simplex virus by granulocytes. J Clin Invest 84: 107–112

Van Strijp JAG, Milternburg LAM, Van Der Tol ME, Van Kessel KPM, Fluit AC, Verhoef J (1990) Degradation of herpes simplex virions by human polymorphonuclear leukocytes and monocytes. J Gen Virol 71: 1205–1209

Van Strijp JAG, Van Der Tol ME, Miltenburg LAM, Van Kessel KPM, Verhoef J (1991) Tumor necrosis factor triggers granulocytes to internalize complement-coated virus particles. Immunology (in press)

Welsh RM, Vargas-Cortes M (1991) Regulation and role of natural killer cells in virus infections. In: Lewis CE, Gee J O'DM (eds) The natural immune system. Oxford University Press, Oxford (in press)

Wu L-X, Anaraki F, Morahan PS, Leary K (1990) Transient expression of virus-specific promoters in mouse resident peritoneal macrophages. J Leukocyte Biol 48: 229–236

Wu L-X, Eisenstein TK, Morahan PS (1991) Effect of herpes simplex virus type 1 infection on cytokine gene expression in activated mouse macrophages. In: Walsh L, Block TM, Crowell RL, Jungkind DL (eds) Innovations in antiviral development and the detection of virus infection. Plenum, New York (in press)

Yasumoto S, Okabe N, Mori R (1986) Role of epidermal Langerhans' cells in resistance to herpes simplex virus infection. Arch Virol 90: 261–271

The Role of Herpes Simplex Virus Glycoproteins in Immune Evasion

G. DUBIN[1], N. O. FISHMAN[1], R. J. EISENBERG[3,4], G. H. COHEN[2,3], and H. M. FRIEDMAN[1,5]

1 Introduction

Viruses have evolved mechanisms that favor their own survival by modifying the effectiveness of the host immune response. Although viral strategies for immune evasion are varied, many members of the herpesvirus family encode proteins that have immunoregulatory functions. Human cytomegalovirus (CMV) encodes a protein that shares homology with the heavy chain of the major histocompatibility complex class I molecule (BECK and BARRELL 1988). This protein interacts with β_2-microglobulin and may interfere with cytotoxic T cell recognition of infected cells (BROWNE et al. 1990). BCRF1, a protein encoded by Epstein-Barr virus (EBV), is a homologue of interleukin-10 and inhibits synthesis of cytokines by activated T helper cells (HSU et al. 1990).

[1] Infectious Diseases Section, Department of Medicine, School of Medicine, University of Pennsylvania
[2] Department of Microbiology and
[3] Center for Oral Health Research, School of Dental Medicine, University of Pennsylvania
[4] Department of Pathobiology, School of Veterinary Medicine, University of Pennsylvania
[5] Department of Pathology and Laboratory Medicine, Children's Hospital of Philadelphia, Philadelphia, Pennsylvania 19104, USA

Current Topics in Microbiology and Immunology, Vol. 179

Herpes simplex virus (HSV) encodes at least nine glycoproteins that are found on the viral envelope and on the surface of infected cells. These glycoproteins mediate essential steps in the viral replicative cycle (attachment, entry, and egress of virus from the infected cell) and are targets for immune attack. Three glycoproteins, gE, gI, and gC, interact with components of the humoral immune system and modulate the immune response to infection. gE and gI form a receptor for the Fc domain of immunoglobulin G (IgG), while gC is a receptor for complement C3 fragments C3b and iC3b. Both Fc receptors (FcRs) and C3 receptors (C3Rs) are found on the virion and on the surface of infected cells. Although gE, gI, and gC are not essential for virus replication in tissue culture, mutants lacking FcR or C3bR activity are rarely isolated from patients, suggesting that these functions are conserved in vivo and are likely important in the pathogenesis of HSV infection. Other herpesviruses express similar immune receptors, including CMV and varicella zoster virus, which encode FcRs (Keller et al. 1976; Litwin et al. 1990), and EBV which encodes a glycoprotein that inhibits activation of the alternative complement pathway (Mold et al. 1988).

2 HSV C3 Receptor

C3 is one of the most versatile and multifunctional molecules of the complement system; it plays a pivotal role in both the classical and alternative pathways of complement activation. Complement receptor type 1 (CR1) is the predominant mammalian C3b binding protein. It protects the host cell by: (1) accelerating the decay of the classical and alternative C3 convertases, enzymes which are important activators of the cascade; (2) blocking the interaction of C3b with C5 to prevent generation of the C5b–C9 membrane attack complex; and (3) acting as a cofactor for the degradation of C3b (reviewed in Lambris 1988). Proteins which interact with C3 have been identified on a variety of microorganisms, including HSV types 1 and 2, EBV vaccinia virus, *Trypanosoma*, *Plasmodium*, and *Schistosoma* species, and *Candida albicans* (reviewed in Dierich et al. 1989). This section will focus on the identification, characterization, and function of the C3 R expressed by HSV; this protein protects both the virus and virus-infected cells from complement-mediated attack.

3 General Properties of the HSV C3 Receptor

Infection with HSV-1 induces the expression of receptors for the Fc portion of IgG and for C3b on cells (Cines et al. 1982). Studies have established that glycoprotein C (gC) is the C3b binding protein (Friedman et al. 1984; Smiley and

FRIEDMAN 1985). Monoclonal antibodies (MAbs) to gC1 block the binding of C3b, while MAbs to other HSV-1 glycoproteins do not. Additionally, a gC-negative mutant of HSV-1 (MP) does not bind C3b; however, cloning the gC gene from the HSV-1 strain F into the MP mutant reconstitutes C3b binding activity. More recently, HSV-1 has been shown to bind another C3 fragment, iC3b (KUBOTA et al. 1987); this binding was mediated by gC (TAL-SINGER et al. 1991).

C3b and iC3b receptors are expressed on cells infected with HSV-1, but not with HSV-2. However, as purified glycoproteins, both gC1 and gC2 bind these C3 fragments (EISENBERG et al. 1987). Cells transiently transfected with the *gC2* gene also bind C3b and iC3b (SEIDEL-DUGAN et al. 1988; TAL-SINGER et al. 1991). Potential explanations for the absence of C3R activity at the HSV-2-infected cell surface include lower affinity of gC2 than gC1 for C3 fragments, inhibition of C3 binding to gC2 by other HSV-2 glycoproteins, or quantitative differences in the amount of gC2 expression. This last possibility is unlikely since experiments demonstrate that, although HSV-2-infected cells express less gC than HSV-1-infected cells, the reduction is only approximately 50% (JENNINGS et al. 1987).

4 Structure of the HSV C3 Receptor (gC)

gC of HSV-1 has a predicted molecular weight of 55 K for the unprocessed polypeptide chain and an actual molecular weight of approximately 120 K by SDS-PAGE; it maps to the U_L*44* open reading frame, which encodes for 511 amino acids (FRINK et al. 1983; MCGEOCH et al. 1988). There are nine potential N-linked glycosylation sites which play an important role in C3bR activity. Treatment of purified gC1 with endo F or endo H does not alter C3b binding, but treatment of purified gC2 with these endoglycosidases significantly decreases binding (EISENBERG et al. 1987). Glycosylation of gC also influences immunoreactivity with MAbs (SJÖBLOM et al. 1987). Neuraminadase treatment of cells infected with HSV-1 markedly enhances the binding of C3b and iC3b to gC, indicating that sialic acid modifies receptor activity (SMILEY and FRIEDMAN 1985; TAL-SINGER et al. 1991).

Initial attempts to identify binding sites on gC for C3b utilized a panel of anti-gC1 MAbs to block receptor activity and gC1 MAb-resistant mutants with single amino acid changes in each of the four antigenic sites of gC1. Several of these mutants failed to bind C3b, but no particular antigenic region of gC1 correlated with C3b binding (FRIEDMAN et al. 1986). In-frame deletion mutants of gC2 were next used to identify C3b binding sites on gC (SEIDEL-DUGAN et al. 1988). One mutant protein lacking residues 26–73 bound C3b on transfected cell surfaces, indicating that this region is not required for receptor activity. Two other deletion mutants did not bind MAbs or C3b. However, these mutations disrupted conformation of the proteins and thus precluded any conclusions with respect to the involvement of these sites in C3b binding.

In-frame linker-insertion mutagenesis was employed to more finely map C3bR domains of gC2 (SEIDEL-DUGAN et al. 1990). Forty-one mutants were constructed, each having single, double, or triple insertions of four amino acids spaced across the protein. Each mutant protein was expressed on the transfected cell surface and bound one or more MAbs recognizing separate discontinuous epitopes, suggesting that each protein was folded properly. Three distinct regions were found to be important for C3b binding, since insertions within these areas abolished receptor activity. Region I is located between amino acids 102 and 107, region II is located between residues 222 and 279, and region III is located between residues 307 and 379. More recently, the domains of gC2 which bind iC3b were mapped and were similar to the C3b binding regions (TAL-SINGER et al. 1991).

5 Role of the HSV C3 Receptor in Immune Evasion

Purified gC1 modulates complement activation by accelerating the decay of C3bBb, the amplification convertase of the alternative complement pathway (FRIES et al. 1986). gC1 also inhibits C5bC6-initiated reactive lysis of C3b-bearing cells, suggesting that gC1 interferes with the interaction of C3b with C5 or C5b. gC2 differs substantially from gC1 when tested for these functions (EISENBERG et al. 1987). Purified gC2 prolongs the survival of the active C3bBb enzyme, rather than accelerating its decay. In contrast to gC1, there is no depression of the hemolytic efficiency of the complement system by gC2, and gC2 has little effect on the late acting complement components. Dissimilarities between gC1 and gC2 may have implications for understanding the possible role of these glycoproteins in modifying the hosts immune response.

gC1 and gC2 protect against complement-mediated viral neutralization (MCNEARNEY et al. 1987; HARRIS et al. 1990); gC1 also plays a protective role at the infected cell surface (HARRIS et al. 1990; HIDAKA et al. 1991). HARRIS and coworkers compared a mutant HSV-1 strain that does not express gC1 at the cell surface and that does not bind C3b with its parental strain. Cells infected with the mutant virus were more susceptible to cytolysis mediated by antibody and complement or complement alone. In addition, cells transfected with the *gC1* gene were protected against complement-mediated lysis. Protection of infected and transfected cells was mediated by inhibition of the alternative complement pathway and was most marked when complement was present in the absence of antibody. This suggests that gC1 may be important during primary HSV infection, at a time when complement is present but antibody has not yet appeared.

Experiments in mice failed to demonstrate a protective role for gC1 or gC2 in pathogenesis; however, these studies were not designed specifically to evaluate the importance of C3 binding (DIX et al. 1983; JOHNSON et al. 1986; SUNSTRUM et al. 1988). SUNSTRUM and coworkers used HSV-1 strain KOS which binds C3 poorly; the mouse strain studied was DBA/2, which is partially deficient in C5, negating a contribution by complement in virus clearance. The route of inoculation was intracerebral, which may bypass any opportunity for complement to effect pathogenesis. Most importantly, gC1 binds poorly to murine C3 fragments (HIDAKA et al. 1991).

6 HSV Fc Receptors

In 1964, WATKINS first demonstrated that HeLa cells infected with HSV adsorbed IgG-sensitized sheep erythrocytes in a rosetting assay; incubation of cells with rabbit anti-HSV serum inhibited this phenomenon (WATKINS 1964). Subsequent studies showed that a variety of cell types acquire the ability to bind free or antigen-associated IgG following infection with HSV-1 or HSV-2 and that binding can be attributed to a receptor for the Fc domain of IgG that is induced on the surface of infected cells (YASUDA and MILGROM 1968; WESTMORELAND and WATKINS 1974; NAKAMURA et al. 1978).

7 General Properties of HSV Fc Receptors

HSV FcRs have been identified on the virion envelope and are expressed at low levels on the surface of cells immediately after exposure to virus, presumably by transfer to the cellular membrane during viral entry (PARA et al. 1980). FcR expression on cells begins to increase as early as 2 h postinfection and reaches a plateau by 8–24 h (WESTMORELAND and WATKINS 1974; NAKAMURA et al. 1978; MCTAGGART et al. 1978). Binding of IgG to HSV FcRs is temperature-dependent, with maximal binding occurring at 37 °C (WESTMORELAND and WATKINS 1974); binding occurs at reduced levels at 4 °C and 15 °C, suggesting that receptor mobility contributes to Fc binding. Electron microscopy has demonstrated that, following binding to the cell membrane, peroxidase-labeled IgG is internalized (COSTA et al. 1978).

Several studies indicate that FcR activity can be modified by proteolytic enzymes. Trypsin treatment of infected cells results in a rapid loss of FcRs, with complete regeneration occurring over 2–8 h (NAKAMURA et al. 1978; MCTAGGART et al. 1978). In contrast, treatment of infected cells with neuraminidase enhances binding of IgG up to twofold (COSTA et al. 1978; MCTAGGART et al. 1978).

8 Fc Receptor Structure

Using IgG affinity chromatography, BAUCKE and SPEAR initially isolated an Fc binding glycoprotein, designated gE, from extracts of HSV-1-infected cells (BAUCKE and SPEAR 1979). Although three electrophoretically distinct Fc binding polypeptides were identified, the different forms were thought to represent sequential stages of posttranslational processing of a single protein product. The fully processed form of gE has a molecular weight of 80 K. Sequencing of the unique short region of the HSV-1 genome has mapped gE to the US8 open reading frame (MCGEOCH et al. 1985). An HSV-2 glycoprotein has been identified that is antigenically related to gE and has Fc binding activity (PARA et al. 1982).

JOHNSON and coworkers demonstrated that gI, the protein product of the US7 open reading frame, coprecipitated with gE using nonimmune IgG or antibodies against gE or gI (JOHNSON and FEENSTRA 1987). gI has a molecular weight of 70 K and is possibly one of the three Fc binding polypeptides originally described by BAUCKE and SPEAR. JOHNSON and coworkers further observed that gE and gI form a complex that constitutes a functional FcR and that neither glycoprotein alone can bind IgG (JOHNSON et al. 1988). Several recent studies, however, show that gE, in the absence of gI, has Fc binding activity, but that its Fc binding activity is enhanced by gI (BELL et al. 1990; DUBIN et al. 1990; HANKE et al. 1990). Cells expressing gE alone bind IgG-coated erythrocytes (that is, IgG aggregates) but little or no soluble IgG (that is, IgG monomers). Cells expressing gE and gI bind a greater amount of IgG-coated erythrocytes than cells expressing gE alone and also bind soluble IgG, including IgG monomers. These observations indicate that HSV encodes two types of FcRs: (1) gE is an FcR for IgG complexes; (2) gE and gI interact to form a higher affinity FcR that in addition binds IgG monomers.

9 Binding Characteristics and Comparison with Other Fc Receptors

Three distinct IgG FcRs have been identified on mammalian cells and are distinguished, in part, by their affinities for monomeric IgG and IgG aggregates or complexes (reviewed in UNKELESS 1989). Mammalian FcRs II and III are low affinity receptors that bind IgG aggregates and complexes but not IgG monomers; these receptors, therefore, have binding characteristics that are analogous to gE. Mammalian FcRI, a high affinity receptor, binds monomeric IgG ($K_a = 10^8–10^9\,M^{-1}$) and is analogous to the FcR formed by gE and gI. Protein A of *Staphylococcus aureus*, a high affinity FcR, similarly binds monomeric IgG. Scatchard analysis of HSV FcRs raises the possibility that two types of FcRs with different binding affinities may be expressed on infected cells and indicates that the K_a for the higher affinity receptor is $2 \times 10^7 M^{-1}$ (JOHANSSON and BLOMBERG 1990).

JOHANSSON and coworkers have shown that the HSV FcR and protein A of *Staphylococcus aureus* demonstrate similar patterns of binding for human IgG subclasses: both bind IgG4 > IgG1 > IgG2, and neither binds IgG3 (JOHANSSON et al. 1984). This differs from mammalian FcRI which binds IgG1 = IgG3 > IgG4, but does not bind IgG2. Furthermore, the HSV FcR has been shown to bind to IgG in a region containing the C_H2 and C_H3 domains, the same domains recognized by *S. aureus* protein A, while mammalian FcRI binds a region in the C_H2 domain of IgG (JOHANSSON et al. 1986). With respect to species specificity, the HSV FcR binds human, rabbit, swine, and sheep IgG but not murine IgG; this species pattern differs from that of *S. aureus* protein A (JOHANSSON et al. 1985).

10 Role of the HSV Fc Receptors in Immune Evasion

Although the role of HSV FcRs in modulating the course of infection in vivo is unknown, several studies suggest that these receptors protect the virus or virus-infected cells from host immune attack by binding nonimmune IgG. DOWLER and VELTRI demonstrated that nonimmune IgG or purified Fc protects HSV-2 from antibody neutralization or thermal inactivation at 37 °C (DOWLER and VELTRI 1984). ADLER and coworkers showed that heat-aggregated nonimmune IgG protects HSV-infected cells from lysis by antibody and complement or sensitized lymphocytes (ADLER et al. 1978). It is proposed that binding of nonimmune IgG sterically hinders access of immune effector elements to the virus or virus-infected cell. Binding of nonimmune IgG or Fc fragments to HSV-infected cells has also been shown to inhibit virus growth and may play a role in the establishment or maintenance of viral latency (COSTA et al. 1977). The mechanism of this inhibition has not been defined, and it remains controversial whether the HSV FcR mediates the responses (JOHANSSON and KJELLEN 1988).

Recently, FRANK and FRIEDMAN demonstrated that the HSV FcR is able to bind anti-HSV IgG by participating in a process called antibody bipolar bridging (FRANK and FRIEDMAN 1989). This occurs when an antiviral IgG molecule binds to an antigenic target by its Fab end and concomitantly to the HSV FcR by its Fc end. It has been postulated that antibody bipolar bridging may protect the virus and virus-infected cells from host immune defenses mediated by the Fc domain of antiviral IgG. FRANK and FRIEDMAN showed that antibody bipolar bridging protects HSV-1 from antibody- and complement-mediated neutralization, presumably by blocking the binding of complement component Clq to the Fc domain of antiviral IgG (FRANK and FRIEDMAN 1989). Recent studies by DUBIN and coworkers demonstrate that antibody bipolar bridging protects HSV-infected cells from antibody-dependent cellular cytotoxicity (DUBIN et al. 1991). In addition, they demonstrated that bipolar bridging of antiviral IgG on infected cells inhibits the binding of Clq. Inhibition of Clq binding may protect cells from lysis by the classical complement pathway. As described elsewhere in this chapter, virus-encoded receptors for C3 fragments protects infected cells from lysis by inhibiting alternative complement pathway activation and inhibiting generation of the membrane attack complex. It is likely, therefore, that viral FcRs and C3Rs function together to protect cells from antibody and complement attack.

References

Adler R, Glorioso J, Cossman J, Levine M (1978) Possible role of Fc receptors on cells infected and transformed by herpesvirus: escape from immune cytolysis. Infect Immun 21: 442–447

Baucke RB, Spear PG (1979) Membrane proteins specified by herpes simplex viruses. V. Identification of an Fc-binding glycoprotein. J Virol 32: 779–789

Beck S, Barrell BG (1988) Human cytomegalovirus encodes a glycoprotein homologous to MHC class I antigens. Nature 331: 269–272

Bell S, Cranage M, Borysiewicz L, Minson T (1990) Induction of immunoglobulin G Fc receptors by recombinant vaccinia viruses expressing glycoproteins E and I of herpes simplex virus type 1. J Virol 64: 2181–2186

Browne, H, Smith G, Beck S, Minson T (1990) A complex between the MHC class I homologue encoded by human cytomegalovirus and $\beta 2$ microglobulin. Nature 347: 770–772

Cines DB, Lyss AP, Bina M, Corkey R, Kefalides NA, Friedman HM (1982) Fc and C3 receptors induced by herpes simplex virus on cultured human endothelial cells. J Clin Invest 69: 123–128

Costa J, Rabson AS, Yee C, Tralka TS (1977) Immunoglobulin binding to herpes virus-induced Fc receptors inhibits virus growth. Nature 269: 251–252

Costa J, Yee C, Nakamura Y, Rabson A (1978) Characteristics of the Fc receptor induced by herpes simplex virus. Intervirology 10: 32–39

Dierich MP, Huemer HP, Prodinger WM (1989) C3 binding proteins of foreign origin. In: Lambris JD (ed) The third component of complement: chemistry and biology. Springer, Berlin Heidelberg New York, pp 163–178 (Current topics in microbiology and immunology, vol 153)

Dix RD, McKendall RR, Baringer JR (1983) Comparative neurovirulence to herpes simplex virus type 1 strains after peripheral or intracerebral inoculation of BALB/c mice. Infect Immun 40: 103–112

Dowler K, Veltri R (1984) In vitro neutralization of HSV-2: inhibition of binding of normal and purified Fc to virion Fc receptor (FcR). J Med Virol 13: 251–259

Dubin G, Frank I, Friedman HM (1990) Herpes simplex virus type 1 encodes two Fc receptors which have different binding characteristics for monomeric immunoglobulin G (IgG) and IgG complexes. J Virol 64: 2725–2731

Dubin G, Socolof E, Frank I, Friedman HM (1991) The herpes simplex virus type 1 Fc receptor protects infected cells from antibody dependent cellular cytotoxicity. J Virol 65: 7046–7050

Eisenberg RJ, Ponce de Leon M, Friedman HM, Fries LF, Frank MM, Hastings JC, Cohen GH (1987) Complement component C3b binds directly to purified glycoprotein C of herpes simplex virus types 1 and 2. Microb Pathog 3: 423–435

Frank I, Friedman HM (1989) A novel function of the herpes simplex virus type 1 Fc receptor: participation in bipolar bridging of antiviral immunoglobulin G. J Virol 63: 4479–4488

Friedman HM, Cohen GH, Eisenberg RJ, Seidel CA, Cines DB (1984) Glycoprotein C of herpes simplex virus 1 acts as a receptor for the C3b complement component on infected cells. Nature 309: 633–635

Friedman HM, Glorioso JC, Cohen GH, Hastings JC, Harris SL, Eisenberg RJ (1986) Binding of complement component C3b to glycoprotein C of herpes simplex virus type 1: mapping of gC-binding sites and demonstration of conserved C3b binding in low-passage clinical isolates. J Virol 60: 470–475

Fries LF, Friedman HM, Cohen GH, Eisenberg RJ, Hammer CH, Frank MM (1986) Glycoprotein C of herpes simplex virus 1 is an inhibitor of the complement cascade. J Immunol 137: 1636–1641

Frink RJ, Eisenberg R, Cohen G, Wagner EK (1983) Detailed analysis of the portion of the herpes simplex virus type 1 genome encoding glycoprotein C. J Virol 45: 634–647

Hanke T, Graham FL, Lulitanand V, Johnson DC (1990) Herpes simplex virus IgG Fc receptors induced using recombinant adenovirus vectors expressing glycoproteins E and I. Virol 177: 437–444

Harris SL, Frank I, Yee A, Cohen GH, Eisenberg RJ, Friedman HM (1990) Glycoprotein C of herpes simplex virus type 1 prevents complement-mediated cell lysis and virus neutralization. J Infect Dis 162: 331–337

Hidaka Y, Sakai Y, Yasushi T, Mori R (1991) Glycoprotein C of herpes simplex virus type 1 is essential for the virus to evade antibody-independent complement-mediated virus inactivation and lysis of virus-infected cells. J Gen Virol 72: 915–921

Hsu D-H, Malefyt RDW, Fiorentino DF, Dang M-H, Vieira P, DeVries J, Spits H, Mosmann TR, Moore KW (1990) Expression of interleukin-10 activity by Epstein-Barr virus protein BCRFI. Science 250: 830–832

Johansson PJH, Blomberg J (1990) Characterization of herpes simplex virus type 1-induced Fc receptor in its interaction with rabbit immunoglobulin G (IgG). Acta Pathol Microbiol Immunol Scand 98: 685–694

Johansson PJH, Kjellen L (1988) Inhibition of herpes simplex virus growth caused by preparations of animal immunoglobulins is not dependent on Fc–Fc receptor interactions. Intervirol 29: 334–338

Johansson PJH, Hallberg T, Oxelius V-A, Grubb A, Blomberg J (1984) Human immunoglobulin class and subclass specificity of Fc receptors induced by herpes simplex virus type 1. J Virol 50: 796–804

Johansson PJH, Myhre EB, Blomberg J (1985) Specificity of Fc receptors induced by herpes simplex virus type 1: comparison of immunoglobulin G from different animal species. J Virol 56: 489–494

Johansson PJH, Schroder AK, Nardella FA, Mannik M, Christensen P (1986) Interaction between herpes simplex type 1-induced Fc receptor and human and rabbit immunoglobulin G (IgG) domains. Immunology 58: 251–255

Jennings SR, Lippe PA, Pauza KJ, Spear PG, Pereira L, Tevethia SS (1987) Kenetics of expression of herpes simplex virus type 1-specific glycoprotein species on the surfaces of infected murine, simian, and human cells: flow cytometric analysis. J Virol 61: 104–112

Johnson DC, Feenstra V (1987) Identification of a novel herpes simplex virus type 1-induced glycoprotein which complexes with gE and binds immunoglobulin. J Virol 61: 2208–2216

Johnson DC, McDermott MR, Chrisp C, Glorioso JC (1986) Pathogenicity in mice of herpes simplex virus type 2 mutants unable to express glycoprotein C. J Virol 58: 36–42

Johnson DC, Frame MC, Ligas MW, Cross AM, Stow ND (1988) Herpes simplex virus immunoglobulin G Fc receptor activity depends on a complex of two viral glycoproteins, gE and gI. J Virol 62: 1347–1354

Keller R, Peitchel R, Goldman JN (1976) An IgG Fc receptor induced in cytomegalovirus-infected human fibroblasts. J Immunol 116: 772–777

Kubota Y, Gaither TA, Cason J, O'Shea JJ, Lawley TJ (1987) Characterization of the C3 receptor induced by herpes simplex virus type I infection of human epidermal, endothelial, and A431 cells. J Immunol 138: 1137–1142

Lambris JD (1988) The multifunctional role of C3, the third component of complement. Immunol Today 9: 387–393

Lehner T, Wilton JMA, Shillitoe EJ (1979) Immunological basis for latency, recurrences, and putative oncogenicity of herpes simplex virus. Lancet ii: 60–62

Litwin V, Sandor M, Grose C (1990) Cell surface expression of the varicella zoster virus glycoproteins and Fc receptor. Virology 178: 263–272

McGeoch DJ, Dolan A, Donald S, Rixon FJ (1985) Sequence determination and genetic content of the short unique region in the genome of herpes simplex virus type 1. J Mol Biol 181: 1–13

McGeoch DJ, Dalrymple MA, Davison AJ, Dolan A, Frame MC, McNab D, Perry LJ, Scott JE, Taylor P (1988) The complete DNA sequence of the long unique region in the genome of herpes simplex virus type 1. J Gen Virol 69: 1531–1574

McNearney TA, Odell C, Holers M, Spear PG, Atkinson JP (1987) Herpes simplex virus glycoproteins gC-1 and gC-2 bind to the third component of complement and provide protection against complement-mediated neutralization of viral infectivity. J Exp Med 166: 1525–1535

McTaggart SP, Burns WH, White DO, Jackson DC (1978) Fc receptors induced by herpes simplex virus. I. Biologic and biochemical properties. J Immunol 121: 726–730

Mold C, Bradt BM, Nemerow GR, Cooper NR (1988) Epstein-Barr virus regulates activation and processing of the third component of complement. J Exp Med 168: 949–969

Nakamura Y, Costa J, Tralka TS, Yee CJ, Rabson AS (1978) Properties of the cell surface Fc-receptor induced by herpes simplex virus. J Immunol 121: 1128–1131
Para MF, Baucke RB, Spear PG (1980) Immunoglobulin G (Fc)-binding receptors on virions of herpes simplex virus type 1 and transfer of these receptors to the cell surface by infection. J Virol 34: 512–520
Para MF, Goldstein L, Spear PG (1982) Similarities and differences in the Fc-binding glycoprotein (gE) of herpes simplex virus types 1 and 2 and tentative mapping of the viral gene for this glycoprotein. J Virol 41: 137–144
Seidel-Dugan C, Ponce de Leon M, Friedman HM, Fries LF, Frank MM, Cohen GH, Eisenberg RJ (1988) C3b receptor activity on transfected cells expressing glycoprotein C of herpes simplex virus types 1 and 2. J Virol 62: 4027–4026
Seidel-Dugan C, Ponce de Leon M, Friedman HM, Eisenberg RJ, Cohen GH (1990) Identification of C3b-binding regions on herpes simplex virus type 2 glycoprotein C. J Virol 64: 1897–1906
Sjöblom I, Lundström M, Sjögren-Jansson E, Glorioso JC, Jeansson S, Olofsson S (1987) Demonstration of highly carbohydrate dependent epitopes in the herpes simplex virus type 1-specified glycoprotein C. J Gen Virol 68: 545–554
Smiley ML, Friedman HM (1985) Binding of complement component C3b to glycoprotein C is modulated by sialic acid on herpes simplex virus type 1-infected cells. J Virol 55: 857–861
Sunstrum JC, Chrisp CE, Levine M, Glorioso JC (1988) Pathogenicity of glycoprotein C negative mutants of herpes simplex virus type 1 for the mouse central nervous system. Virus Res 11: 17–32
Tal-Singer R, Seidel-Duagn C, Fries LF, Huemer HP, Eisenberg RJ, Cohen GH, Friedman HM (1991) Herpes simplex virus glycoprotein C is a receptor for complement component iC3b. J Infect Dis 164: 750–753
Unkeless JC (1989) Function and heterogeneity of human Fc receptors for immunoglobulin G. J Clin Invest 83: 355–361
Watkins JF (1964) Adsorption of sensitized sheep erythrocytes to HeLa cells infected with herpes simplex virus. Nature 202: 1364–1365
Westmoreland D, Watkins JF (1974) The IgG receptor induced by herpes simplex virus: studies using radioiodinated IgG. J Gen Virol 24: 167–178
Yasuda J, Milgrom (1968) Hemadsorption by herpes simplex-infected cell cultures. Int Arch Allergy 33: 151–170

Immunopathology of Herpes Simplex Virus Infections

M. Z. DOYMAZ and B. T. ROUSE

1 Introduction

The great majority of accounts of immune mechanisms in virus infections have dealt with protective aspects, but it is evident that immune reactions against infected cells or viral antigens may cause tissue damage. Indeed, it is likely that, during clearance of most if not all viruses, an inflammatory reaction ensues after a T cell response is generated (MIMS 1983). It is perhaps a matter of semantics as to whether or not this should be described as immunopathology (IP). For the purpose of this review, IP will be considered as a chronic inflammatory reaction mediated primarily by the immune system that results in tissue damage or a change in immune function which is initiated and perhaps perpetuated by virus infections. Such situations result more commonly following infections with noncytopathic viruses which persist in an expressed form in the body (SISSONS and BORYSIEWICZ 1985). Although herpes simplex virus (HSV) is persistent, clearly it is not usually expressed, and productive infections are highly cytopathic. Consequently, HSV is unlikely to participate in many immunopathological situations. Before discussing these, our review briefly surveys the state of affairs with other agents which best indicate immunopathological disease mechanisms

Department of Microbiology, College of Veterinary Medicine University of Tennessee, Knoxville, Tennessee 37996-0845, USA

Current Topics in Microbiology and Immunology, Vol. 179

and which provide clues to understanding those processes that occur in HSV-induced IP.

Immunopathological states result from many circumstances. Most investigations deal with inflammatory reactions which involve IgE, whose reaction with antigens leads to the elaboration and secretion of inflammatory molecules from mast cells. There are few, if any, well studied situations in which IgE antibodies to viral antigens have been recorded and considered to participate in IP. However, reports of IgE antibodies against HSV have occurred (IDA et al. 1983), but their relevance as regards immunity or IP are not resolved. A second well recognized immunopathological mechanism involves the chronic production of immune complexes that become trapped in tissues and trigger inflammatory responses in which complement components and neutrophils play a major role (MEHTA et al. 1977; OLDSTONE et al. 1983). The critical parameter in the mediation of immune complex disease appears to be the chronic production of low affinity nonneutralizing antibody to some product of the virus (MIMS 1983). Thus, the immune response fails to dislodge the virus and the entrapment of complexes in various tissues triggers inflammatory responses. Immune complex disease, at least with viral antigens, has received little recent study and it is unclear if this mechanism accounts for the IP of any HSV infections. However, as discussed subsequently, possible candidates include stromal keratitis and the rare disease erythema multiforme (ROUSE 1985).

2 T Cell-Mediated IP

Most current interest in viral-induced IP involves T cell-mediated mechanisms and the prospect that many viruses may in some way trigger autoimmune inflammatory responses which usually involve T cells (TER MEULEN 1989). The latter topic received an excellent review in a recent volume in this series (OLDSTONE 1989). Especially prominent among hypotheses to explain autoimmune reactions triggered by virus infections is that the viruses could share antigens with self-tissues (molecular mimicry). The immune response to these molecular mimics may in some way disrupt the immunoregulatory process which normally prevents autoreactive responses from occurring. Exactly how molecular mimicry on the part of viral antigens leads to autoimmunity is not understood but has been discussed in some detail in the aforementioned volume (FUJINAMI 1989).

2.1 $CD8^+$ T Cell-Mediated IP

The role of T cells and the likely means by which they contribute to IP in viral infections has received the most study with lymphocytic choriomeningitis (LCM) viral infection of mice (BUCHMEIER et al. 1980). In this disease, a fatal immunopathological disease occurs around 1 week following intracerebral infection of immunologically competent mice. The disease is characterized by massive extravasation of lymphocytes and monocytes into the cerebrospinal fluid (CSF) with the inflammatory process involving the leptomeninges, choroid plexus, and ependyma. These sites represent the distribution of infected cells. The key to disease development is the $CD8^+$ subset of T cells which react with limited epitopes on the LCM virus (ALLAN et al. 1987; DIXON et al. 1987). From in vitro studies, a nonapetide appears to be the dominant epitope recognized in one strain of mice and this recognition is effected by a very limited T cell repertoire (WHITTON et al. 1989; KLAVINSKIS et al. 1989; PIRCHER et al. 1987; YANAGI et al. 1990). The in vivo studies clearly indicate an essential role for $CD8^+$ cells. In fact, the inflammatory exudates contain few if any $CD4^+$ cells (DOHERTY et al. 1990). Depleting $CD8^+$ but not $CD4^+$ cells ablates disease and the lesions are only transferable with $CD8^+$ cells. Indeed, the disease can be induced by intracerebral adoptive transfers of $CD8^+$ LCM-specific clones (BEAZINGER et al. 1986). Since $CD8^+$ cells are well known to mediate cytotoxicity in vitro, it could be that the inflammation represents a direct T cell destruction of virus-infected endothelial cells with the subsequent escape of blood elements into the CSF. However, only very few of the $CD8^+$ cells appear to be virus-specific and the frequency of the cytotoxic T lymphocyte (CTL) percursors in the CSF is only of the order of 1 in 2000 (CEREDIG et al. 1989). A more likely mechanism of $CD8^+$-mediated inflammation is the release of various cytokines which recruit and retain both nonspecific $CD8^+$ T cells and monocytes (DOHERTY et al. 1990). This delayed-type hypersensitivity (DTH) reaction is more typically the function of $CD4^+$ cells and appears to be the case in HSV infections (ROUSE 1985).

Another model of $CD8^+$ T lymphocyte-mediated IP is hepatitis B virus (HBV) infection. This virus is usually not cytotoxic to liver cells and the damage during acute hepatitis is assumed to result from the immune response against the virus (HOLLINGER 1990). Intrahepatic $CD8^+$ T cells from biopsy specimens of patients and the presence of HBV capsid antigen-specific CTLs in the circulation lend support for the notion that $CD8^+$ cells are primary mediators of the IP (MONDELLI et al. 1990). This hypothesis is difficult to prove in the human setting, but it recently received considerable support with the development of a transgenic mouse model (MORIYAMA et al. 1990). In this model, HBV envelope protein (env) expression was accomplished in hepatocytes of transgenic mice. Adoptive transfer of HBV env (amino acids 20–41)-specific $CD8^+$ T cells into syngeneic transgenic mice induced hepatocyte lysis as assessed by histopathological and liver function assays. In vitro-propagated, monoclonal, $CD8^+$ T cells were used in adoptive transfer experiments. Thus, these results not only identified the immunopathological T cell subset but also the epitope recognized by the cells.

2.2 $CD4^+$ T Cell-Mediated IP

In addition to participating in effector aspects of immunity, $CD4^+$ cells play an essential role in regulating the immune response (BRAAKMAN et al. 1987). Immunopathological situations can result either from a sustained reaction of $CD4^+$

antigen-specific cells against infected targets or from the indirect effect of virus infection on CD4$^+$ cells themselves. This causes changes in their production of the many cytokines which serve to regulate immunity. There is a growing awareness that this latter mechanism may be an important cause of disease (MIYAJIMA et al. 1988), although viral effects on macrophage and B cell function may be an even more common cause of immune dyscrasia (MORAHAN et al. 1985). For instance, some lesions noted in the Epstein-Barr herpesvirus (EBV) infection are associated with polyclonal activation of B lymphocytes which produce antibodies, including autoantibodies, which in turn can mediate an inflammatory response (RICKINSON et al. 1985). Although HSV can replicate in macrophages, no cytokine-mediated dyscrasia has been reported to result from such interactions.

The reaction of CD4$^+$ effector cells with virus-infected targets usually gives rise to a DTH reaction. Thus CD4$^+$ effector cells, although they can be cytotoxic (BRAAKMAN et al. 1987), more usually mediate effector function by the secretion of cytokines, which serve to recruit and change the physiological state of other cell types. This normally protective response is not judged as immunopathological unless prolonged, occurs in some critical location, or sets off some chronic, tissue damaging, autoreactive response. One of the better studied situations in which the CD4$^+$ response is judged to be immunopathological is infection with Theiler's murine encephalitis virus (TMEV), a picornavirus (CLATCH et al. 1985). This natural pathogen of mice causes an enteric infection and occasionally paralysis. Some strains cause a biphasic disease of the central nervous system when injected intracranially (LIPTON 1975). The early disease appears to result from direct effects of the virus on motor neurons in the spinal cord. The later disease has been used as a model for multiple sclerosis since it results in an inflammatory demyelinating lesion that is confined to the white matter (RODRIGUEZ et al. 1986). This is thought to result from virus persisting in oligodendroglia cells combined with a CD4$^+$-mediated T cell response against them (CLATCH et al. 1986). Depletion of CD4$^+$ cells prior to the demyelinating stage results in reduced incidence and disease severity. Thus, the disease appears to be mediated by CD4$^+$ T cells (WELSH et al. 1987). Contradicting these results, RODRIGUEZ and SRIRAM (1988), using the same T subset depletion approach, reported that CD8$^+$ T cells were the primary mediators of demyelination in TMEV-infected mice. Thus, it is possible that CD4$^+$ T cell involvement in the IP is not direct but may proceed by providing help to CD8$^+$ T cells. Similar autoreactive, presumably CD4$^+$-mediated, responses occur in coronavirus-induced demyelination in both mice and rats (SORENSON et al. 1980; OUBINE et al. 1988). In such animals, lymphocytes react with myelin basic protein, and cells can transfer the disease in both its clinical and pathological expression to naive animals (WATANABE et al. 1983). Contributing to the pathology in these instances could be the induction of class II antigen expression on nerve cells, which occurs because of the release of interferon-γ (IFN-γ) by CD4$^+$ lymphocytes (MASSA et al. 1986). Such class II-positive cells might act as presenters of autoantigens to induce an autoreactive myelin basic protein-specific response.

It is clearly evident from the work of MOSSMAN and others that $CD4^+$ lymphocytes in both mice and humans are divisible into at least two principal functional subgroups based upon their cytokine secretion patterns (MOSSMANN and COFFMAN 1989). The two subsets, Th1, comprised of cells that secrete interleukin-2 (IL-2) and IFN-γ and Th2, comprised of cells that secrete IL-4 and IL-5, play differing roles in immune regulation and presumably in immunity. The best evidence for the varying role of Th1 and Th2 lymphocytes in immunity comes from studies of parasite resistance in mice. In mice strains in which the response in dominated by Th1 cells, animals are protected from infection, but if the Th2 cell response predominates, animals develop lesions that are considered immunopathological (LIEW 1987). The route of exposure to antigen also affects the balance of the response, with intravenous and subcutaneous exposure favoring a Th1 and Th2 response, respectively (LIEW et al. 1985). Factors which influence Th1 or Th2 induction remain ill-defined but may reflect antigen presentation differences, the relative abundance of T cells of unique repertoire, and other regulatory influences on the Th1: Th2 ratio. It does appear as if differing antigenic epitopes can trigger one or the other subset. Thus, an octamer from protein p183 of *Leishmania major* is a preferential inducer of Th2 in BALB/c mice and immunization with this peptide leads to a heightened susceptibility to disease (LIEW et al. 1990). It is becoming evident that the different cytokines produced by Th1 and Th2 might not only influence the outcome of infection, but cytokines from one subset may influence the function of the other. Currently, there is much interest in the observation that IL-10, produced by Th2 cells in vitro, has a down-regulatory effect on Th1 cells (FIORENTINO et al. 1989). In this regard, IL-10 acts as an antigen-nonspecific suppressor factor.

Although the correlation between Th1 and protection and Th2 and IP appears to hold true in certain parasitic infections, it is uncertain exactly how these effects are carried out. More than likely, however, Th2 cytokines, such as IL-4, which is a potent modulator of the IgE response, are mechanistically involved (PLAUT 1990). The IgE could, in turn, mediate a Jones-Mote-type hypersensitivity reaction.

3 IP in HSV Infections

It is characteristic of HSV that, after infection, virus multiplies locally, reaches peak titers between the second and sixth days, and usually becomes undetectable after the tenth day. The local reaction is accompanied by an inflammatory response which, at least in the initial phases, probably represents a reaction to the cells killed by virus replication. The later phases of the inflammatory response are likely to be in part immune-mediated, since T cell reactions can be detectable as early as 4 days after primary infection. During recrudescent infections, when

T cells can be restimulated in 2–3 days from their memory precursors, the inflammatory response may be largely immune-mediated. If the T cell immune system is functioning normally, clearance of virus rapidly occurs and extensive or prolonged tissue injury does not result from the T cell antiviral response. Accordingly, the state of affairs is usually not considered as immunopathological. However, if: (a) inflammatory responses occurring in certain locations are prolonged; (b) the regenerative capacity of cells is destroyed or limited, or (c) the repair deposits material that impairs the function of some organs, then the situation can be considered as immunopathological. Herpetic infections of the eye best illustrate this scenario, e.g., herpetic stromal keratitis (HSK), which as discussed below appears to largely represent an immunopathological response to HSV in the cornea.

Uveitis is a second HSV-induced ocular lesion that may have an immunopathological explanation. This is a very rare syndrome in humans, but a rabbit animal model has been established to study its pathogenesis (OH 1976). Animals are infected via the cornea or anterior chamber and those animals with lesions in the uveal tract are given a second intravitreal infection of virus. The prolonged inflammatory response which follows is considered, at least in part, immunopathological (OH et al. 1985). Accordingly, the lesion can also be induced with replication incompetent virus and efforts to recover infectious virus from the inflammed uvea are usually unsuccessful (OH 1976). The actual mechanism of the IP has not been fully investigated. Conceivably the virus could set off an autoimmune reaction to some component in the uveal tract. A similar situation may be occurring in human uveitis. Thus, uveitis can be established by immunizing animals against as S antigen isolated from the retina (RAO et al. 1989), and uveitis patients respond to management with corticosteroids (DINNING 1984).

3.1 Herpetic Stromal Keratitis

Herpetic infections of the cornea are the leading nontraumatic cause of blindness in the United States and there are an estimated 500000 individuals who suffer from herpetic eye disease (HYNDIUK and GLASSER 1986). Although acute infections of the corneal surface are well controlled by antivirals, especially acycloguanosine, the corneal stroma may also become involved, particularly as a result of recurring infections associated with reactivation from latency (O'BRIEN 1986). This inflammatory disease in the stroma often leads to corneal scarring and neovascularization, permanent endothelial destruction with corneal edema, and even secondary glaucoma and cataract (KAUFMAN 1978). One reason for suspecting that the stromal manifestations represent an immunopathological disease is that they respond well to corticosteroid immunosuppression (KAUFMAN 1978). There is a strong, yet still unproven, suspicion that HSK represents an inflammatory immunogenic response to intrastromal viral antigens (METCALF and KAUFMAN 1976). Epithelial keratitis, on the other hand, which occurs both as a primary and a recurring infection, is a mainfestation of viral-induced cytolysis of the superficial epithelium and responds well to antivirals (DANIELS

1982). In this form of herpetic corneal disease, the use of corticosteroids leads to further ocular damage. In humans there are two principal expressions of HSK—a nonnecrotizing (disciform keratitis) and a necrotizing variety. In both instances neither the role of replicating HSV nor the part played by the immune response in contributing to the lesion is understood. It is also unclear what determines the pattern of recovery and resolution.

Most of our mechanistic understanding of the pathogenesis of HSK comes from studies in rabbit and mouse models (METCALF et al. 1976; O'BRIEN 1984; OPREMCAK et al. 1990). Of the two, the rabbit is the most convenient to model the histopathology and investigate the effects of treatment, but it is the mouse model which permits a better understanding of the role of the immune response in contributing to the tissue damage. In the rabbit, the disease usually takes the necrotizing form and in this expression polymorphonuclear neutrophils (PMNs) dominate the inflammatory responses (MEYERS-ELLIOT and CHITJIAN 1981). Indeed, the severity of the disease is markedly suppressed if PMNs are depleted or complement activity is inhibited (MEYERS-ELLIOT and CHITJIAN 1981). Consequently, the mechanism of pathology may be a combination of direct viral damage perhaps combined with immune complex-mediated IP (ROUSE 1985). However, the issue has not been resolved and the rabbit is not a convenient model to resolve immunological controversies, particularly if T cells are involved.

Fortunately, mice can also be experimentally infected with certain strains of HSV and develop HSK; however, abrasion of the cornea before infection is usually required and only some strains of mice are susceptible (STULTING et al. 1985). Thus, the C57/BL and C3H strains are resistant, whereas HSK can readily be induced in BALB/c and A/J mice. Murine HSK is a model of disciform nonnecrotizing keratitis and in this form the inflammation is dominated, after an initial stage, by lymphocytes and other mononuclear cells or by PMNs, depending upon the viral strain used (HENRICKS and TUMPEY 1990; OPPEMCAK et al. 1990). Replicating virus, or viral antigens, are difficult to demonstrate. It is the murine model which we and others have used in an attempt to better understand the IP of HSK (METCALF et al. 1979; RUSSELL et al. 1984). The aim has been to establish the immune processes which contribute to the tissue damage and hopefully to identify the target viral antigens and the role of viral replication in the disease expression. The mouse is not an ideal model to simulate human HSK since the latter usually results from repeated recurrent infections. HSK is usually a progressive disease in mice and spontaneous recurrences from latent infection do not occur. When such recurrences are induced experimentally by procedures such as iontophoresis the mice do not clinically express the disease (O'BRIEN 1984).

3.1.1 T Cells in the IP of HSK

Early histopathological studies of HSK by METCALF and HELMSEN (1977) led to the suspicion that HSK represents an immunopathological reaction. Thus, mononuclear inflammatory cells late in the disease were found in close contact with keratinocytes, which themselves appeared to be undergoing degenerative changes. The lesions were interpreted to represent T cell-mediated attacks on infected cells. Support for this idea was later dramatically presented by METCALF et al. (1979) when they demonstrated that HSK lesions fail to occur in athymic mice. Such mice, however, developed severe corneal epithelial infections and virus disseminated beyond the ocular site of infection which included the brain. Mice ultimately developed encephalomyelitis.

RUSSELL et al. (1984) confirmed METCALF'S findings and extended them using reconstitution experiments. Thus, athymic mice given adoptive transfers of HSV immune cells rapidly developed a severe form of HSK. The peak reaction occurred earlier in athymic mice that received adoptive transfers of immune cells than following primary infection of euthymic mice. The cells transferring the keratitis were identified as T cells, but their subset nature or function was not definitively characterized. On the basis of these indirect approaches it was concluded that CTLs were principally involved. To understand the mechanisms of IP, METCALF (1984) compared the IP-restorative capacity of lymphocytes taken from mice previously infected either subcutaneously or intraperitoneally with HSV. The former population was found to be more effective, but the reasons were not elucidated. We now know, however, that populations from subcutaneously immunized mice would have far greater DTH mediating activity than populations from mice immunized intra-peritoneally (NASH and GELL 1983). This may argue a role for the involvement of DTH.

Most groups working on the murine HSK model seem to agree that the lesion largely represents T cell-mediated IP, but the actual mechanism remains uncertain. Basically, there are two schools of thought. Either the lesion primarily represents a $CD8^+$-mediated immunopathological response, with in vivo cytotoxicity playing a principal part, or the lesion represents a $CD4^+$-mediated one. In this instance, the $CD4^+$ cells are assumed to release various cytokines, including destructive ones such as tumor necrosis factor, or they themselves mediate cytotoxicity.

A primary role for $CD8^+$ CTLs has been advanced by the observations of HENDRICKS and colleagues (1989a,b). They took advantage of the peculiar phenomenon in the eye that, following injection of an antigen into the anterior chamber (AC), animals develop a partial state of immune tolerance which primarily affects aspects of cell-mediated immunity (KAPLAN and STREILEIN 1977; KSANDER and HENDRICKS 1987). Using this model, 20 min after the AC injection, the virus was applied topically to the corneas. The expectation was that the animals would be unable to generate some aspects of cellular immunity and consequently would not develop HSK in the topically infected eye. This, in fact, was found to be the case. To implicate $CD8^+$ CTLs as the probable mediators of HSK, HENDRICKS et al. (1989b) later exploited the observation that certain gC mutant viruses, when injected into the AC, tolerize mice only for CTLs and not for DTH. Thus, if CTLs were the mediators of IP, then AC injection of the mutant was also expected to ablate HSK following topical infection of wild-type virus. Such results were obtained. This rather indirect argument has not been confirmed by others, and it is curious to note that the usual pattern of events following AC exposure to antigen is reported to be somewhat different. Accordingly, AC injection of virus tolerizes mice for DTH but not for CTLs (STREILEIN 1987; IGIETSEME et al. 1989). Thus, it is uncertain why the initial experiments of the HENDRICKS group would not result in HSK, if indeed CTLs were involved, because in that model the CTL response should still be operative.

The notion that $CD4^+$ T cells mediate IP in the murine HSK model was advocated by our group and others (NEWELL et al. 1989a, b; HENDRICKS and TUMPEY 1990). Initial evidence was obtained by using monoclonal antibodies specific for T cell subsets to in vivo deplete animals of a particular cell type (NEWELL et al. 1989a). The outcome following corneal infections with the RE strain of HSV to groups of control, $CD4^+$-depleted, or $CD8^+$ T cell-depleted mice were measured. The results were clear-cut. In the absence of $CD4^+$ T cells, HSK lesions were markedly reduced or inapparent. Interestingly, the elimination of $CD8^+$ T cells resulted in more severe HSK and spread of virus to the brain to produce encephalomyelitis. Presumably, $CD8^+$ T cells subserve a protective role against HSK and do not mediate IP. Recently, HENDRICKS and TUMPEY (1990) have confirmed that $CD4^+$ T cells mediate IP in HSK, but only in the same RE HSV strain as was used in our study. Interestingly, they found that IP was mediated by $CD8^+$ T cells if the KOS strain of HSV was used as challenge virus. Furthermore, the nature of stromal inflammation induced by the RE strain of HSV-1 seems somewhat different from HSK induced by KOS HSV-1 (HENDRICKS, personal communication). Accordingly, in RE strain-induced inflammation, the migration of Langerhans' cells to the central cornea is more prominent and the inflammation is characterized as neutrophilic, whereas in KOS-induced HSK mononuclear cells are prominent. Presently, no adequate explanation has been put forward as to why the strain of the virus induces markedly different disease patterns Antigen differences between the viral strains could provide the answer to this question.

To further resolve the part played by $CD4^+$ and $CD8^+$ T cells in HSK IP, we have established an experimental model in which thymectomized and T cell-depleted (T(-)) mice were reconstituted with different numbers and types of T cells from immune or naive mice (DOYMAZ and ROUSE 1991). Thymectomy combined with the administration of anti-CD4 and anti-CD8 sera provided a model animal that failed to develop HSK corneal challenge. Donor cells for adoptive transfer were taken from normal Balb/c mice undergoing HSK. Accordingly, retropharyngeal and cervical lymph node cells were stimulated in vitro with HSV for 4 days then negatively depleted in vitro of $CD4^+$ or $CD8^+$ prior to transfer to T(−) mice which had been infected on the cornea with HSV 2 days previously. We observed that HSK occurred only in recipients of unfractionated and $CD8^+$-depleted populations but not in recipients of the $CD4^+$-depleted population. Moreover, if sufficient cells were transferred (2×10^7 or higher) a disease comparable in severity to that observed in the direct HSK model occurred. Naive T lymphocytes, irrespective of the surface phenotype, were unable to induce HSK upon transfer to T(−) mice. Taken together, our results provide further evidence that $CD4^+$ but not $CD8^+$ T cells were mediating the IP in HSK, a notion recently confirmed by others (HENDRICKS and TUMPEY 1990).

3.1.2 Mechanism of T Cell-Mediated IP in HSK

Although there is reasonable agreement that T cells of some type are involved in mediating IP in HSK, how this occurs is far from clear. Indeed since a particular T cell subset can subserve several different effector mechanisms, the studies related to these questions are more difficult and the answers may not be clear-cut (DOHERTY et al.1990). Added uncertainties in the case of HSK come from the fact that assays measuring the functions of T cells in HSV mice models usually require in vitro stimulation steps, increasing the possibility of introduction of artifacts. Furthermore, the microanatomy of corneal tissues makes in situ analysis of the function of T cells almost impossible (STREILEIN 1987). Thus, T cells obtained from local lymph nodes (retropharyngeal and cervical) seem to be the best candidate population to study. Since, in vivo depletion and adoptive transfer experiments indicated that $CD4^+$ T cells were crucially involved in HSK IP, we concentrated our efforts on the function of these cells.

One of the more universally accepted functions of $CD4^+$ T cells is the mediation of DTH reactions. This notion has laso been clearly demonstrated for HSV in mice (NASH et al. 1981; ROUSE 1985). Furthermore, we have data indicating that, both in normal or reconstituted mice experiencing HSK, a significant DTH response is mounted against HSV antigens. Histopathological examination of diseased corneas shows that inflammatory cell infiltrates are reminiscent of conventional DTH reactions. Another approach to assess the role of DTH in immune responses is to determine the lymphokine secretion patterns of $CD4^+$ cells (MOSSMANN and COFFMAN 1989). One expects that cytokines such as IL-2 and IFN-γ would be involved and that modulating the levels of such cytokines would influence the outcome of HSK. That this is indeed the case has been noted recently by HENRICKS (personal communication) and in our own studies (DOYMAZ and ROUSE, unpublished).

Direct cytotoxicity on virally infected corneal cells is another mechanism by which $CD4^+$ cells might participate in HSK IP. This type of CTL predominates in human HSV infections (SCHMID 1988; YASUKAWA et al. 1988). Moreover, keratinocytes readily express MHC class II molecules upon stimulation by IFN-γ and can present HSV antigens to autologous $CD4^+$ T cells (CUNNINGHAM and NOBLE 1989). The elements required for $CD4^+$ CTL recognition, such as $CD4^+$ T cells themselves and class II-expressing antigen-presenting cells, are found in murine eyes with HSK (MCBRIDE et al.1988; DOYMAZ et al.1991). Furthermore, we have recently documented that murine $CD4^+$ cells may express HSV-specific class II-restricted cytotoxicity (KOLAITIS et al. 1990). Similar $CD4^+$ cytotoxic cells can be demonstrated in draining retropharyngeal and cervical lymph nodes of mice experiencing HSK (Table 1). These cells specifically lysed syngeneic target cells infected with HSV. Moreover $I\text{-}A^+$ syngeneic target cell lysis was still intact in $CD8^+$ T cell-depleted mice (DOYMAZ et al. 1991). These findings were particularly interesting in the context of HSK IP, since mice lacking $CD8^+$ lymphocytes experience a more pronounced stromal disease than do normal animals (NEWELL et al. 1989a). We also demonstrated the presence of $CD4^+$ T lymphocytes and

Table 1. Retropharyngeal and cervical lymph nodes of Balb/c mice experiencing herpetic stromal kerotitis contain a distinct T cell population which is $CD4^+$ and lyses MHC class III-bearing syngeneic target cells but not cells expressing class I only[a]

Target cells	A.20-Mock	A.20-HSV	A.20-Vaccinia	EMT.6-Mock	EMT.6-HSV	L-HSV
Experiment I						
c′ only	11 (+/−1)[b]	33 (+/−1)	5 (+/−3)	11 (+/−1)	56 (+/−3)	9 (+/−)
Anti-Thy 1.2	0 (+/−1)	0 (+/−1)	0 (+/−1)	0 (+/−1)	4 (+/−1)	2 (+/−2)
Anti-CD4	3 (+/−2)	17 (+/−1)	1 (+/−1)	6 (+/−4)	52 (+/−5)	7 (+/−2)
Anti-CD8	6 (+/−1)	22 (+/−1)	3 (+/−2)	9 (+/−1)	9 (+/−1)	7 (+/−1)
Experiment II						
c′ only	12 (+/−1)	54 (+/−1)	9 (+/−2)	7 (+/−1)	47 (+/−3)	11(+/−1)
Anti-CD4	5 (+/−1)	49 (+/−3)	12 (+/−3)	8 (+/−1)	47 (+/−2)	4 (+/−1)
Anti-CD8	4 (+/−1)	23 (+/−1)	7 (+/−1)	8 (+/−1)	12 (+/−2)	5 (+/−2)

[a] Mice were challenged on the cornea with 1×10^6 $TCID_{50}$ RE strain HSV-1, and 10 days later lymph node cells were obtained. After 3 days culture in the presence of UV-inactivated HSV, lymphocytes were depleted with monoclonal plus complement treatment as indicated

[b] Specific lysis was determined on syngeneic A.20 (H-2^d, I-A^d) and EMT.6 (H-2^d) cells and on allogeneic L (H-2^k) cells by a 4 h^{51} chromium release assay, % lysis only at 50:1. Effector cells target cells ratio is presented

MHC II-expressing potential antigen-presenting cells in the corneal stroma during the clinical phase of the disease (DOYMAZ et al. 1991). It will be important to determine if $CD4^+$ class II-restricted CTLs are present in corneal tissues and the cytokine secretion profile of such cells.

3.1.3 The Role of Non-T Cell-Mediated Mechanisms in HSK

It remains probable that the IP of clinical HSK is not confined to T cell mechanisms. Indeed, FOSTER and colleagues (1986) have indicated a possible role for humoral immune-mediated pathology. Thus, in studies of disease patterns in immunoglobulin heavy chain gene (IgH) congenic mice strains, susceptibility to HSK was correlated with the IgH phenotype. Such studies indicate that the immunoglobulin repertoire may in some way influence susceptibility to HSK but how this occurs remains ill-defined. It is possible that the type of antibody response made by susceptible strains renders them more likely to generate inflammatory immune complexes. Indeed, in the rabbit model of HSK, the disease pattern may be consistent with immune complex-mediated IP (ROUSE 1985). Thus, in such animals PMNs are prominently involved in the inflammatory response (MEYERS-ELLIOT and CHITJIAN 1981) and plasma cells can be found in the limbus and corneal tissues in some instances (METCALF et al. 1976; MEYERS-ELLIOT and CHITJIAN 1981). Moreover, polyclonal anti-PMN antibody treatment or complement depletion of rabbits significantly reduced the severity of the disease (MEYERS-ELLIOT and CHITJIAN 1981). In the rabbit model, it is difficult to assess any additional role of T cells. Morever, it is also possible that the anti-PMN serum used for depletion lacked strict specificity and was inhibitory to T cells. Accordingly, the diminished HSK and reduction in anti-HSV antibody observed could have resulted from effects on helper T cells (MEYERS-ELLIOT and CHITJIAN 1981).

4 Summary and Prospects

There is good agreement that HSK, in both its natural setting and in animal models, represents an immunopathological response to HSV. We do not know the nature of target antigens or how and in which cells they are expressed, nor indeed are we certain about the actual mechanism(s) involved in IP. With the recent availability of new animal model systems, along with an ever increasing spectrum of antisera that can ablate various cell types, T cell receptor expression, and levels of cytokines in vivo, rapid progress in further defining the mechanisms of IP in HSK is to be anticipated. Hopefully, the outcome of such studies will be the development of new treatment or prevention strategies for this tragic disease.

Acknowledgment. This work was supported by Public Health Services Grants EY 05093 and AI 14981.

References

Allan JE, Dixon JE, Doherty PC (1987) Nature of inflammatory process in the central nervous system of mice infected with lymphocytic choriomeningitis virus. In: Oldstone MB (ed) Arenaviruses. Biology and immunotherapy. Springer, Berlin Heidelberg New York, pp 131–143 (Current topics in microbiology and immunology, vol 134)

Beazinger J, Hengartner H, Zinkernagel RM (1986) Cloned cytotoxic T cells specific for lymphocytic choriomeningitis virus induce acute disease and primary footpad swelling in infected mice. Med Microbiol Immunol (Berl) 175: 201

Braakman E, Rottevell FTM, van Bleek G, van Seventer GA, Lucas CJ (1987) Are MHC Class-II restricted cytotoxic T lymphocytes important. Immunol Today 8: 265–267

Buchmeier MJ, Welsh RM, Dutko FJ, Oldstone MBA (1980) The virology and immunobiology of lymphocytic choriomeningitis virus infection. Adv Immunol 30: 275–331

Ceredig R, Allan JE, Tabi Z, Lynch F, Doherty PC (1989) Phenotypic analysis of the inflammatory exudate in murine lymphocytic choriomeningitis. J. Exp Med 165: 1539–1551

Clatch RJ, Melvold RW, Miller SD, Lipton HL (1985) Theiler's murine encephalomyelitis virus (TMEV) induced demyelinating disease in mice is influenced by the H-2 D region. Correlation with TMEV specific delayed type hypersensitivity reactions. J Immunol 135: 1408–1414

Clatch RJ, Lipton HL, Miller SD (1986) Characterization of Theiler's murine encephalomyelitis virus (TMEV) specific delayed type hypersensitivity responses in TMEV-induced demyelinating disease: correlation with clinical signs. J Immunol 136: 920–927

Cunningham AL, Nobel JR (1989) Role of keratinocytes in human recurrent herpetic lesions. Ability to present herpes simplex virus antigen and act as targets for T lymphocyte cytotoxicity in vitro. J Clin invest 83: 490–496

Daniels CA (1982) Viral and rickettsial infections. In: Gardner A, Klintworth GK (eds) Pathobiology of ocular diseases. A dynamic approach. Dekker, New York, pp 249–266

Dinning WJ (1984) The therapy of uveitis. A review. In: Saari KM (ed) Uveitis updated. Excerpta Medica, New York, pp 477–481

Dixon JE, Allan JE, Doherty PC (1987) The acute inflammatory process in murine lymphocytic choriomeningitis virus is dependent on Ly 2^+ immune T cells. Cell Immunol 107: 8–14

Doherty PC, Allan JE, Lynch F, Ceredig R (1990) Dissection of an inflammatory process induced by $CD8^+$ T cells. Immunol Today 11: 55–59

Doymaz MZ, Rouse BT (1991) Herpetic stromal keratitis: an immunopathological disease mediated by $CD4^+$ T lymphocytes (to be published)

Doymaz MZ, Mercadal CM, Destephano D, Rouse BT (1991) MHC-II restricted, $CD4^+$ cytotoxic T lymphocytes (CTL) specific for herpes simplex virus-1: implications for the development of herpetic stromal keratitis (HSK) in mice (to be published)

Fiorentino DF, Bond BW, Mossmann TR (1989) Two types of mouse T helper cells. IV. T H2 clones secrete a factor that inhibits cytokine production by T HI cells. J Exp Med 170: 2081–2095

Foster CS, Tsai Y, Monroe JG, Campbell R, Cestari M, Wetzig R, Knipe D, Greene MI (1986) Genetic studies on murine susceptibility to herpes simplex keratitis. Clin Immunol Immunopathol 40: 313–325

Fujinami RS (1989) Immune responses against myelin basic protein and/or galactocerebroside cross-react with viruses: implications for demyelinating disease. In: Oldstone MB (ed) Molecular mimicry. Cross-reactivity between microbes and host proteins as a cause of auto immunity. Springer, Berlin Heidelberg New York, pp 93–100 (Current topics in microbiology and immunology, vol 145)

Hendricks RL, Tumpey T (1990) Contribution of virus and immune factors to herpes simplex virus type 1 induced corneal pathology. Invest Opthalmol Vis Sci 31: 1929–1939

Hendricks RL, Epstein RJ, Tumpey T (1989a) The effects of cellular immune tolerance to HSV-1 antigens on the immunopathology of HSV-1 keratitis. Invest Opthalmol Vis Sci 30: 105–115

Hendricks RL, Tao MSP, Glorioso JC (1989b) Alterations in the antigenic structure of two major HSV-1 glycoproteins, gC and gB, influence immune regulation and susceptibility to murine herpes keratitis. J Immunol 142: 263–269

Hollinger FB (1990) Hepatitis B virus. In: Fields BN, Knipe DM (eds) Virology, 2nd edn. Raven, New York, pp 2171–2224

Hyndiuk RA, Glasser DB (1986) Herpes simplex keratitis. In: Tabbarra K, Hyndiuk RA (eds) Infections of the eye. Diagnosis and management. Little Brown, Boston, pp 343–368

Ida S, Siragenian RP, Notkins AL (1983) Cell bound and circulating IgE antibody to herpes simplex virus. J Gen Virol 64: 533–537

Igietseme JV, Calzada PJ, Gonzalez AR, Streilein, Atherton SS (1989) Protection of mice from herpes simplex virus-induced retinitis by in vitro activated immune cells. J Virol 63: 4808–4813

Kaplan HJ, Streilein JW (1977) Immune response to immunization via the anterior chamber of the eye. I. F1 lymphocyte induced immune deviation. J Immunol 118: 809–814

Kaufman HE (1978) Herpetic keratitis. Invest Opthalmol Vis Sci 117: 941–957

Klavinskis LS, Lindsay-Whitton J, Oldstone MBA (1989) Molecularly engineered vaccine which express an immunodominant T cell epitope induces cytotoxic T lymphocytes that confer protection from lethal virus infection. J Virol 63: 4311–4316

Kolaitis G, Doymaz M, Rouse BT (1990) Demonstration of MHC class II restricted cytotoxic T lymphocytes in mice against herpes simplex virus. Immunology 71: 101–106

Ksander BR, Hendricks RL (1987) Cell-mediated immune tolerance to HSV-1 antigens associated with reduced susceptibility to HSV-1 corneal lesions. Invest Ophthalmol Vis Sci 28: 1986–1992

Liew FY (1987) Analysis of host protective and disease promoting T cells. Ann Inst Pasteur Immunol 138: 749–755

Liew FY, Hale C, Howard JG (1985) Prophylactic immunization against experimental Leishmaniasis. IV subcutaneous immunization prevents the induction of protective immunity against fatal *Leishmania major* infection. J Immunol 135: 2095–2101

Liew FY, Millott SM, Schmidt JA (1990) A repetitive peptide of leishmania can activate T helper type 2 cells and enhance disease progression. J Exp Med 172: 1359–1365

Lipton HL (1975) Theiler's virus infection in mice: an unusual biphasic disease process leading to demyelination. Infect Immun 11: 1147–1155

Massq PT, Dorries R, Ter Muelen V (1986) Viral particles induce Ia expression on astrocytes. Nature 320: 543–546

McBride BW, McGill JI, Smith JL (1988) MHC-class-I and class-II antigen expression in normal human corneas and in corneas from cases of herpetic keratitis. Immunology 65: 583–587

Mehta PD, Kane A, Thormar H (1977) Quantitation of measles virus-specific immunoglobulins in serum, CSF and brain extract from patients with subacute sclerosing panencephalitis. J Immunol 118: 2254–2261

Metcalf JF (1984) Herpetic keratitis in normal and athymic mice. In: Tabbarra K, Cello RM (eds) Animal models of ocular diseases. Thomas, Spring field, pp 53–62

Metcalf JF, Helmsen R (1977) Immunoelectron microscopic localization of herpes simplex virus antigens in rabbit cornea with anti-human IgG-antiferrin hybrid antibodies. Invest Ophthalmol Vis Sci 9: 779–786

Metcalf JF, Kaufman HE (1976) Herpetic stromal keratitis: evidence for cell-mediated immunopathogenesis. Am Ophthalmol 82: 827–837

Metcalf JF, McNeil JI, Kaufman HE (1976) Experimental disciform edema and necrotizing keratitis in rabbit. Invest Ophthalmol Vis Sci 12: 979–985

Metcalf JF, Hamilton DS, Reichert RW (1979) Herpetic keratitis in athymic (nude) mice. Infect Immun 26: 1164–1171

Meyers-Elliot RH, Chitjian PA (1981) Immunopathogenesis of corneal inflammation in herpes simplex virus stromal keratitis: role of the polymorphonuclear leucocytes. Invest Ophthalmol Vis Sci 20: 784–798

Mims CA (1983) Immunopathology in virus diseases. Philos Trans R Soc Lond 308: 18–198

Miyajima A, Miyataka S, Schreurs J, DeVries J, Arai N, Yokota T, Arai K (1988) Coordinate regulation of immune and inflammatory responses by T cell derived lymphokines. FASEB J 2: 2462–2473

Mondelli MV, Chisari FV, Ferrari C (1990) The cellular immune response to nucleocapsid antigens in hepatitis virus infections. Springer Semin Immunopathol 12: 25–31

Morahan PS, Connor JR, Leary KR (1985) Viruses and versatile macrophage. Br Med Bull 41: 15–21

Moriyama T, Guilhot S, Klopchin K, Moss B, Pinkert CA, Palmiter RD, Brinster RI, Kanagawa O, Chisari FV (1990) Immunobiology and pathogenesis of hepatocellular injury in hepatitis B virus transgenic mice. Science 248: 361–364

Mossmann TR, Coffman RL (1989) Heterogeneity and cytokine secretion patterns and functions of helper T cells. Adv Immunol 46: 111–147

Nash AA, Gell PGH (1983) Membrane phenotype or murine effector and suppressor T cells involved in delayed type hypersensitivity and protective immunity to herpes simplex virus. Cell Immunol 75: 348–355

Nash AA, Gell PGH, Wildy P (1981) Tolerance and immunity in mice infected with herpes simplex virus. Simultaneous induction of protective immunity and tolerance to delayed-type hypersensitivity. Immunology 43: 153–159
Newell CK, Martin S, Sendele D, Mercadal CM, Rouse BT (1989a) Herpes simplex virus-induced stromal keratitis: role of T-lymphocytes subsets in immunopathology. J Virol 63: 769–775
Newell CK, Sendele D, Rouse BT (1989b) Effects of $CD4^+$ and $CD8^+$ T lymphocyte depletion on the induction and expression of herpes simplex stromal keratitis. Regul Immunol 2: 366–369
O'Brien WJ (1984) Herpetic eye diseases in animals as models for therapeutic studies of acute and latent herpesvirus infections. In: Rapp F, Herpesvirus. Liss, New York, pp 101–120
O'Brien WJ (1986) Antiviral agents. In: Tabbarra K, Hyndiuk R (eds) Infections of the eye. Little Brown, Boston, pp 257–274
Oh JO (1976) Primary and secondary herpes simplex uveitis in rabbits. Surv Ophthalmol 21: 178–182
Oh JO, Minasi P, Grabner G, Ohashi Y (1985) Suppression of secondary herpes simplex uvetis by cyclosporine. Invest Ophthalmol Vis Sci 26: 494–500
Oldstone MBA (1989) Moleculear mimicry as a mechanism for the cause and as a probe uncovering etiological agent(s) of autoimmune disease. In: Oldstone MB (ed) Molecular mimicry. Cross-reactivity between microbes and host proteins as a cause of autoimmunity. Springer, Berlin Heidelberg New York, pp 127–135 (Current topics in microbiology and immunology, Vol 145)
Oldstone MBA, Tishon A, Buchmeier MJ (1983) Virus induced immune complex disease: genetic control of Clq binding complexes in the circulation of mice persistently infected with lymphocytic choriomeningitis virus. J Immunol 130: 912–918
Opremcak EM, Rice BA, Wells PA, Foster CS (1990) Histology and immunohistology of Igh-1 restricted herpes simplex keratitis in Balb/c congenic mice. Invest Ophthalmol Vis Sci 31: 305–312
Oubina JR, Carballal G, Laguens RP, Quintans C, Merani S, Weisshenbacher MC (1988) Mortality induced by adoptive immunity in Junin virus-infected athymic mice Intervirology 29: 61–67
Pircher H, Michalopoulos EE, Iwamato A, Ohashi PS, Beazinger J, Hengartner H, Zinkernagel RM, Mak T (1987) Molecular analysis of the antigen receptor of virus-specific cytotoxic T cells and identification of a new V-alpha family. Eur J Immunol 17: 1843–1847
Plaut M (1990) Antigen specific lymphokine secretory patterns in atopic diseases. J Immunol 144: 4497–4500
Rao NA, Atalla L, Linker-Israeli M, Chen FY, George IV FW, Martin WJ, Steinman L (1989) Suppression of experimental uveítis in rats by anti-IA antibodies. Invest Ophthalmol Vis Sci 30: 2348–2355
Rickinson AB, Tao QY, Wallace LE (1985) The Epstein-Barr virus as a model of virus-host interactions. Br Med Bull 41: 75–79
Rodriguez M, Sriram S (1988) Successful therapy of Theiler's virus-induced demyelination (DA strain) with monoclonal anti-Ly 2 antibody. J Immunol 140: 2950–2955
Rodriquez M, Pease LR, David CS (1986) Immune-mediated injury of virus-infected oligodendrocytes. Immunol Today 7: 359–363
Rouse BT (1985) Immunopathology of herpesvirus infections. In: Roizman B, Lopez C (eds) Herpesviruses, vol 4. Plenum, New York, pp 103–119
Russell RG, Nasisse MP, Larsen HS, Rouse BT (1984) Role of T lymphocytes in the pathogenesis of herpetic stromal keratitis. Invest Ophthalmol Vis Sci 25: 938–944
Schmid DS (1988) The human MHC-II restricted cellular response to herpes simplex virus type-1 is mediated by $CD4^+$ $CD8^+$ T cells and is restricted to the DR region of MHC complex. J Immunol 140: 3610–3616
Sissons JGP, Borysiewiczs LK (1985) Viral immunopathology. Br Med Bull 41: 34–40
Sorenson O, Percy D, Dales S (1980) In vivo and in vitro models of demyelinating diseases. III. JHM virus infection of rats. Arch Neurol 37: 478–484
Streilein JW (1987) Immune regulation and the eye: a dangerous compromise. FASEB J 1: 199–208
Stulting RD, Kindle JC, Nahmias AJ (1985) Patterns of herpes simplex keratitis in inbred mice. Invest Ophthalmol Vis Sci 26: 1360–1367
Ter Meulen V (1989) Virus-induced, cell-mediated autoimmunity. In: Notkins AL, Oldstone MBA (eds) Concepts in Viral pathogenesis III. Springer, Berlin Heidelberg New York, pp 297–303
Watanabe R, Webe H, TerMeulen V (1983) Adoptive transfer of EAE-like lesions by MBP stimulated lymphocytes from rats with corona virus-induced demyelinating encephalomyelitis. Nature 305: 150–153
Welsh RJC, Tonks P, Nash AA, Blakemore WF (1987) Effects of L3T4 T cell depletion on the pathogenesis of Theiler's murine encephalomyelitis virus infection in CBA mice. J Gen Viral 68: 1659–1667

Whitton J, Tishon A, Lewicki H, Gebhard J, Cook T, Salvato M, Joly E, Oldstone MBA (1989) Molecular analysis of a five amino acid cytotoxic T lymphocyte (CTL) epitope: an immunodominant region which induces non-reciprocal CTL crossreactivity. J Virol 63: 4303–4311

Yanagi Y, Maekawa R, Cook T, Kanagawa O, Oldstone MBA (1990) Restricted V segment usage in T cell receptors from cytotoxic T lymphocytes specific for a major epitope of lymphocytic choriomeningitis virus. J Virol 64: 5919

Yasukawa M, Inatsu A, Kobayashi Y (1988) Helper activity in antigen specific antibody production mediated by $CD4^+$ human cytotoxic T cell clones directed against herpes simplex virus. J Immunol 140: 3419–3425

Contemporary Approaches to Vaccination Against Herpes Simplex Virus

R. L. BURKE

1 Introduction

The task of writing a comprehensive review on the subject of herpes simplex virus (HSV) vaccines is daunting. Instead, this review will cover selected points of recent interest. Others have reviewed the topic more broadly (MEIGNIER 1985; HALL and KATRAK 1986; DIX 1987; STANBERRY 1991) or dealt with selected aspects such as recombinant vaccines (WATSON and ENQUIST 1985) or subunit HSV vaccines (STANBERRY 1990).

Infections caused by HSV constitute extremely prevalent communicable diseases of humans and there is clearly a need for vaccines to control them.

Chiron Corporation, 4560 Horton Street, Emeryville, CA 94608, USA

Current Topics in Microbiology and Immunology, Vol. 179

The two viruses, HSV-1 and HSV-2, cross-react quite extensively (SPEAR 1984), but it is unlikely that a single immunogen would suffice against both agents. A survey of the prevalence of HSV-1 and HSV-2 antibodies reveals that a plateau is reached by age 40–50 years with roughly a 60%–80% and a 20% overall incidence, respectively (JOHNSON et al. 1989). Based on these population distributions, a vaccine directed against HSV-1 would have to be administered as an early childhood immunization, whereas an HSV-2-specific vaccine could be administered to adolescents. The majority of a population that might receive an HSV-2 vaccine would be expected to be HSV-1-seropositive and thus partially primed and partially protected at the time of vaccination. This stratification of risk of acquisition based on serological status is an important consideration in designing appropriate clinical trials to evaluate the efficacy of HSV vaccines. Between 50% and 80% of HSV-2 infections may be unrecognized, seroconversions generating a large pool of individuals with inapparent or asymptomatic infection and leading to increased risk of transmission. In fact, transmission may predominantly occur from source partners who are unaware of their infectious potential. In addition, the high prevalence of HSV-2 infection and the frequency of asymptomatic transmission (discussed below) reveals that ultimately the only way to impact incidence of the disease is by prophylactic vaccination.

The fact that herpes virus causes a primary disease and then establishes a latent infection very rapidly following the acute phase offers a unique and extremely difficult challenge to the goal of developing an HSV vaccine. Traditional vaccine strategies have been very successfully employed against viruses that can cause acute diseases (such as polio, measles, mumps) and even against a virus that can cause persistent disease (hepatitis B), but they have yet to be developed against a virus that establishes a latent state with the potential for reactivation. Since the virus has two phases of infection, acute and recurrent, it may be possible to intervene with vaccine(s) at each point, prophylactically prior to the initial infection event and immunotherapeutically after a latent infection has already been established. Given the perceived difficulty of developing an HSV vaccine coupled with an analysis of the total annual cost of the burden of HSV disease in the United States, the Committee on Issues and Priorities for New Vaccine Development mandated by the Institute of Medicine (1985) stated that "the best that [an HSV] vaccine could achieve would be a 50% reduction in the number of symptomatic primary infections, a 75% reduction in the number of recurrences and a reduction in the severity of episodes averaging about 60%." However, a vaccine with even this limited efficacy would reduce the total cost of the disease burden by 75%. This discussion raises some questions about the feasibility of development of HSV vaccines. For prophylactic application, the optimum vaccine would not only reduce disease, but would also block infection, that is block the establishment of a latent state. To achieve this goal a sufficiently high level of immunity would have to be generated by vaccination to limit virus replication to the site of initial infection and to below a theoretical threshold level required to establish a latent and reactivatable state. Is the concept of a threshold level of virus required to establish a latent infection valid, as some animal studies have suggested (BONNEAU and JENNINGS 1988), and if so, what is this threshold? What is the time window of opportunity to block the establishment of a latent infection? Does the frequency of HSV-specific B and T cells have to be maintained at a high level or is there sufficient time for memory responses to contribute to protection? Can a vaccine derived from components of one virus type generate cross protection or will a heterotype vaccine combining antigens from both HSV-1 and HSV-2 be required for this purpose? From the commercial viewpoint, the rational target is to develop an HSV-2 vaccine. However, this goal may not be the optimum endpoint from the public health point of view.

The development of a therapeutic vaccine to limit recurrent disease manifestations is conceptually a far more difficult goal than is prophylaxis, due to the absence of any prior examples of success. In theory, immunointervention cannot alter the pattern of viral reactivation in the ganglia nor the process of transport of the virus along the axon from the neuronal cell body to the epithelial surface, since these sites are not subject to immunological surveillance. However, if productive infection of epithelial cells is limited, then symptomatic recurrences could be reduced or blocked. As will be discussed more fully below, many HSV vaccine formulations have been tested for therapeutic efficacy in humans with no successes to date. However, promising results have been obtained in a guinea pig model of recurrent genital herpes, in which the frequency and severity of recurrent disease has been reduced by 40%–80% using more recent vaccine formulations (STANBERRY et al. 1988; HO et al. 1990). In those cases, the relative efficacy was dependent upon the details of the immunization, including adjuvant and antigen employed and route and timing of immunization.

2 Approaches to Vaccine Development

2.1 Overview

Traditionally, vaccines have been based upon one of two forms of the infectious agent: (1) a live virus whose virulence or pathogenicity had been attenuated or (2) an inactivated or killed virus preparation. Modern approaches to vaccine development can still be sorted into these two broad categories or combinations of the two, although the techniques employed are more sophisticated and the distinctions may begin to overlap.

2.2 Live Virus Vaccines

The use of live virus vaccines has several advantages compared to the alternative of killed virus vaccines: (1) The immune response induced is, in general, more durable. (2) This approach can elicit a broader immune response, including the induction of MHC class I-restricted cytotoxic T lymphocytes. (3) The vaccination regimen may be simpler, requiring only one or rarely two administrations. (4) Manufacturing may be easier since smaller and fewer doses would be required and no immunoenhancing agent or adjuvant would be required. The main disadvantages arise from the concerns of safety and the necessity to attenuate the virus without reducing its replicative capacity and thus immunogenicity. The stability of the attenuated phenotype during both in vitro production and in vivo amplification must be demonstrated. The theoretical possibility of recombination with wild-type virus should also be addressed. JAVIER et al. (1986) have shown that mixed infection with two complementing, attenuated isolates resulted in in vivo complementation and the production of virulent progeny virus.

Avirulent viruses traditionally were generated by culture and multiple blind passages of the virus in vivo in an alternate host and/or in vitro. A most recent example of the successful application of this approach for herpes viruses is the development of the Oka varicella zoster virus vaccine (ARBETER et al. 1986). If viral virulence factors are well defined then it should be possible to mitigate pathogenicity by rational design by deleting or inactivating the relevant genes. For the case of HSV such factors include those genes or DNA sequences involved in neurotropism, neurovirulence, the establishment and maintenance of latency or reactivation, and cell culture transformation. Achieving this goal requires a precise understanding of the mechanisms of HSV pathogenesis.

Live HSV virus vaccines have been developed and tested in animals models by several groups (KIT 1982; MCDERMOTT et al. 1985; THOMPSON et al. 1986). The most carefully characterized and viable live virus vaccine candidate is an intertypic recombinant developed by ROIZMAN and coworkers (MEIGNIER et al. 1987b, 1988), with extensive animal testing conducted in collaboration with MEIGNIER and coworkers at the Institut Merieux (Charbonnieres les Bains, France) (MEIGNIER et al. 1988, 1990). Two prototype attenuated live virus vaccines, designated R7017 and R7020, have been constructed. Both strains were derived from HSV-1 strain F. Each contains a 700 bp deletion of the coding sequences of the thymidine kinase (*tk*) gene and a second 13 kb deletion extending from the end of the immediate early *α27* gene in the long unique segment of the genome to the promoter regulatory domain of the immediate early *α4* gene in the internal inverted repeat segment. Since the deletion includes most of the internal inverted repeat and those sequences involved in the isomerization/inversion of the genome, the viruses are frozen in the prototype orientation. In addition, a region from the short unique segment containing the HSV-2 genes encoding glycoprotein D (gD), gG, gI, and part of gE was inserted into the area of the large deletion. The two viruses differ in that R7020 also has the *tk* gene reinserted into the larger deletion region. The viruses have been tested in various animal models including mouse intracerebral and eye inoculations, guinea pig intravaginal inoculation, rabbit eye inoculation and aotus (owl monkey) intravaginal and eye inoculation (MEIGNIER et al. 1987b, 1988, 1990). In all of these animals models, the virulence of the recombinant viruses is reduced by at least 10000-fold compared to wild-type virus. In addition, the severity of the acute viral infection is also dramatically attenuated. The viruses are genetically and phenotypically stable following multiple passages both in tissue culture and in animals. Although the viruses can establish latent infections, their ability to due so (as measured by in vitro cocultivation) is substantially reduced compared to wild-type virus. Mice inoculated intracerebrally or in the footpad with the recombinant viruses were protected against an intracerebral challenge inoculation with wild-type virus in a dose-dependent manner (MEIGNIER et al. 1988), and the aotus monkey, a species extremely susceptible to HSV infection, was completely protected against lethal challenge (MEIGNIER et al. 1990). The carefully conducted animal studies have clearly established the attenuation and reduced neurovirulence of these recombinant viruses. However, this attenuation is accompanied by reduced immunogenicity; the recombinant viruses are 10–100 times less immunogenic than the parent HSV-1 (F) strain. The attenuated live virus vaccine has been tested for safety in small scale, human clinical trials conducted by Institut Merieux. The design of the trial, the local and systemic reactions associated with immunization, and the immune responses generated have not yet been reported. There are plans to initiate clinical trials in the United States in the near future. One question that arises with regard to the use of such intertypic recombinant virus vaccines is the difficulty of discerning the immune response generated by vaccination from that engendered by superinfection with wild-type virus. For the present candidate, the antibody response to type-specific epitopes of gB2 or gC2 could serve as markers to detect HSV-2 infection; however, similar markers to detect HSV-1 infection are not readily apparent.

A modification of the live virus vaccine approach is the insertion of the gene(s) encoding protective antigen(s) into an alternative live virus vector or host. Several groups have shown that recombinant vaccinia viruses expressing gD (CREMER et al. 1985; PAOLETTI et al. 1984), gB (CANTIN et al. 1983), or gC (WEIR et al. 1989) can protect mice from lethal challenge infection and reduce the severity of disease in guinea pigs (WACHSMAN et al. 1987). The T cell subsets induced by vaccination with vaccinia viruses expressing herpes proteins and their role in the protective immune response have been investigated for gD (MARTIN and ROUSE 1987; MARTIN et al. 1987) and gB (MCLAUGHLIN-TAYLOR et al. 1988; MARTIN et al. 1989). Six recombinant vaccinia viruses, separately expressing gB, gD, gE, gH, and gI, were tested in the mouse to compare the relative protective response to each glycoprotein. (BLACKLAWS et al. 1990). Vaccinia viruses expressing gB (VacgB) or gD (VacgD) induced HSV neutralizing antibody, increased the rate of virus clearance from the inoculation site, and blocked the establishment of latency. VacgE afforded partial protection, whereas no protective response was generated by viruses expressing gG, gH, or gI. This result is somewhat surprising since well characterized monoclonal antibodies to gH are neutralizing (GOMPELS and MINSON 1986; FULLER et al. 1989), and immunization with VacgG can produce a neutralizing antibody response (SULLIVAN and SMITH 1987). For the case of VacgH, the HSV protein is not expressed on the surface of infected cells (FORRESTER et al. 1991). The gB gene has also been inserted into an adenovirus vector and the recombinant virus protected mice from a lethal challenge infection after a single inoculation (MCDERMOTT et al. 1989). These adenovirus vectors have the intriguing potential of oral delivery to stimulate mucosal immunity. Additional live virus vectors have been employed for other infectious diseases including polio (BURKE et al. 1988) and adeno-associated virus (LEBKOWSKI et al. 1988). The insertion of genes into select bacterial species including *Salmonella*, bacille Calmette-Guerin (BCG), and *Listeria monocytogenes* has been suggested as a particularly effective way to elicit a mucosal immune response (CLEMENTS 1989; SNAPPER et al. 1988). The merit of these possibilities has not been tested for HSV.

2.3 Killed Virus Preparations

The use of killed virus preparations as vaccines for many infectious diseases has a long and successful history. The principal advantage of this approach is the high level of safety attendant with the removal of all replicative and infectious capacity of the preparation. To achieve this level of safety, however, there are several disadvantages: (1) In general, the vaccine-induced immune response is not as persistent at that generated by a live virus inoculation. (2) The immune response may not be as broad since the elicitation of a class I-restricted cytotoxic T lymphocyte (CTL) response is theoretically not possible. (3) The killed virus alone may be poorly immunogenic and may require the addition of an immunopotentiating agent or adjuvant, thus adding to the complexity of the manufacturing process.

2.3.1 Purified Fractions of Virus Proteins

The killed virus approach to vaccine preparation may be refined by using partially or completely purified fractions of the virus particle. Numerous examples evaluating the prophylactic efficacy of crude antigen preparations from virally infected cells, in animal models have been reported (reviewed in MEIGNIER 1985; STANBERRY 1990). Most of these preparations have been detergent extracts enriched in membrane-associated proteins or glycoprotein fractions. These preparations have been tested for efficacy in a variety of animal species including mice, rabbits, guinea pigs, and monkeys. In general, such vaccines have provided good protection against acute disease and death and, in some instances, have been reduced the frequency of latent infection. The vaccines were immunogenic, eliciting multifunctional antibody responses and T cell responses. In some cases the relative efficacy was dependent upon the addition of adjuvants, including alum (ZWEERINK et al. 1981), complete Freund's (KUTINOVA et al. 1980), or PICLC, to potentiate immunogenecity. (KLEIN et al. 1981) Correlations of efficacy with cell-mediated immunity (KUTINOVA et al. 1980) and the antibody response (HILFENHAUS et al. 1981) were observed. Some of these vaccines have also been tested in human trials as described below. Many groups have prepared and tested "subunit" vaccines comprised of individual proteins purified from virus-infected cells. These efforts have focused entirely on using the glycoproteins as immunogens.

The viruses encode nine or more glycoproteins, designated gB, gC, gD, gE, gG, gH, gI, gK and gL and found inserted into the virion envelope and the membrane of the infected cell. In humans there are brisk antibody responses to many of the glycoproteins (EBERLE et al. 1985; ASHLEY et al. 1985a). Several researchers have purified glycoproteins gB, gC, and gD from virus-infected cell lysates using immunoaffinity chromatography and have tested these primarily in a mouse lethal challenge model (reviewed in DIX 1987; STANBERRY 1990). All three of the glycoproteins provided protection against acute disease for a homologous challenge but no or partial protection against a heterologous challenge, with the exception of gD (DIX and MILLS 1985). MEIGNIER et al. (1987a) tested an artificial mixture of gB, gC, gD, and gE purified from virus-infected cells and adsorbed to alum in mice, guinea pigs, and owl monkeys. Mice were well protected against a low dose lethal challenge infection but were less well protected against a high dose lethal challenge. The protective efficacy waned significantly as the time interval between immunization and challenge increased. The vaccine moderated the severity of acute genital herpes in guinea pigs but did not reduce the incidence of infection, the replication of virus in the vagina, or the incidence of latent infection, and no protection was demonstrated in owl monkeys. Neutralizing antibody responses were detectable but were modest and transient in all three species. The difference between this study and those noted above is surprising and may relate to the animal models employed.

2.3.2 Subunit Preparations

The use of recombinant DNA technology for antigen production offers the advantages of safety due to the complete absence of infectious virus or viral nucleic acid, reduced batch to batch variability, and greater yield with reduced production costs. The first application of recombinant DNA technology to HSV vaccine production involved the expression of gD derivative as a fusion protein in *E. coli* (WATSON et al. 1982). Recombinant gD produced in *E. coli* protects mice from lethal infection (WATSON and ENQUIST 1985); gD produced in yeast reduces the severity of acute infection in guinea pigs (STANBERRY et al. 1987); gD produced in mammalian cells protects mice and guinea pigs from acute disease and mortality (BERMAN et al. 1988; SANCHEZ-PESCADOR et al. 1988b); and gD produced in baculovirus protects mice (KRISHNA et al. 1989). Likewise, gB produced in yeast moderates acute disease in guinea pigs (KINO et al. 1989) and, when produced in mammalian cells, completely protects mice and guinea pigs from acute disease (BURKE et al. 1989b). An important point about this technology is that glycoproteins produced in *E. coli* and intracellularly in yeast are not glycosylated and are denatured, whereas those proteins produced in mammalian cells, if carefully handled during purification, can retain their native conformation and are glycosylated. The maintenance of a native structure appears to be essential to generate high levels of neutralizing antibody and protective efficacy (BURKE 1991). To date, the antigenic component of all HSV subunit vaccines has been confined to the glycoproteins. Although the glycoproteins are the predominant target of neutralizing antibody, they may not be the predominant target of cellular immune responses. There is intriguing yet preliminary evidence that the immediate early genes and internal virion proteins may be frequently recognized by cytotoxic effector cells (MARTIN et al. 1990; TIGGES et al. 1991).

2.3.3 Adjuvant Formulation

Many researchers have shown that the inclusion of an adjuvant is also crucial to generate strong immune responses and good levels of protective immunogenicity for subunit vaccines. Adjuvants that have been employed include the traditional alum, the unacceptable complete Freund's adjuvant (CFA), and immune stimulating complexes (ISCOMS) (STANBERRY et al. 1989; ERTURK et al. 1989). We have tested muramyl dipeptides and dipeptide derivatives as adjuvants in a variety of vaccine vehicles including high oil (50%) water-in-oil emulsions, low oil (5%–10%) oil-in-water emulsions, liposomes, and covalently attached to the antigen (SANCHEZ-PESCADOR et al. 1988a; HO et al. 1989). These studies have revealed that subunit vaccines composed of glycoproteins gB and/or gD, combined with a lipophilic derivative of a muramyl tripeptide, MTP-PE, in a low oil emulsion vehicle, are highly immunogenic in a variety of animal species including mice, guinea pigs, rabbits, goats, and baboons. These vaccines elicit antibody responses that are 3- to 15-fold greater than those obtained with the use of alum as adjuvant and are within two-fold of the titer obtained by the use of CFA (BURKE 1991). In the guinea pig model of genital herpes, vaccination with either glycoprotein or the combination vaccine affords near complete protection from viral challenge (BURKE et al. 1989b) and reduces the frequency of the establishment of latency (BURKE et al. 1991). In collaboration with STANBERRY and HO (unpublished observations), these vaccines were demonstrated to have therapeutic efficacy, reducing recurrent disease in previously infected guinea pigs by 40%–80%. The therapeutic efficacy was variable depending upon the antigen and dose employed (STANBERRY and BURKE, unpublished observations), the adjuvant (STANBERRY et al. 1989) and vehicle formulation (HO et al. 1990), and the timing of the intervention (STANBERRY et al. 1989). The validity of this animal model remains to be demonstrated in human clinical trials in which, to mimic the guinea pig model, humans would be vaccinated shortly after the resolution of their primary infection. The use of an HSV-2 vaccine formulation that was therapeutically efficacious in the guinea pig recurrent disease model failed to alter asymptomatic virus shedding in the rabbit eye model of recurrent HSV-1 disease (WECHSLER and NESBURN, personal communication).

2.3.4 Peptide Preparations

The simplification of an HSV vaccine formulation is carried to an elegant point by the use of synthetic peptides themselves as the antigen. The first 23 residues of gD contain two overlapping, continuous B cell epitopes (reviewed in MUGGERIDGE et al. 1990) and a T cell epitope (HEBER-KATZ et al. 1988). This peptide and smaller derivatives have been coupled to Keyhole limpet hemocyanin (KLH) and used to immunize rabbits and mice (EISENBERG et al. 1985). A neutralizing antibody response is produced in rabbits and a protective response in mice. The response was type-specific or type-common depending upon the particular peptide tested (EISENBERG et al. 1985; HEBER-KATZ et al. 1988). As with subunit vaccines, the addition of adjuvants to peptide-based vaccines has been shown to substantially increase their immunogenicity (BRYNESTAD et al. 1990; GEERLINGS et al. 1989).

A variation of the use of peptides as vaccines is to embed the peptide in a carrier particle. The feasibility of this approach has been demonstrated by the insertion of a segment of the *gD1* gene into the pre-S region of the hepatitis B surface antigen (HBsAg). Chimeric particles were produced in yeast containing both gD and HBsAg determinants (VALENZUELA et al. 1985). However, when these particles were used to immunize guinea pigs, no antibody to gD was detected (VALENZUELA, personal communication). Other carrier particles have been suggested including the Ty particles of yeast (ADAMS et al. 1987) and rotavirus particles (FRENCHICK et al. 1989).

3 Testing of HSV Vaccines in Humans

3.1 Early HSV Vaccine Trials in Humans

Very early attempts to protect humans from HSV infection utilized some unusual and, in retrospect, naive approaches. One of these was the use of vaccinia virus, which was proported to generate cross-protective immune responses to HSV and which was tested in several open clinical trials, conducted from 1928–1937, with apparently encouraging results (FREUND 1928; FOSTER and ABSHIER 1937). In a second similar approach, that of boosting immune responses in a nonspecific fashion, it was claimed that immunization with BCG also elicited cross-protective immunity to HSV (BIERMAN 1976; ANDERSON et al. 1974). In this case, the claim has been disproven by conduct of an appropriately controlled human clinical trial (DOUGLAS et al. 1985).

In addition to the use of nonspecific immune enhancers, various extracts of virus-infected cells have been prepared from diverse sources for use as inactivated virus vaccines, including formalized extracts of infected guinea pig footpads (BRAIN 1936) and infected rabbit brain (FRANK 1938). Both of these were reported to have therapeutic efficacy in small open trials. The preparations were later refined by transferring the production system from animals to the amnion of chicken embroys (ANDERSON et al. 1950). Institutionalized children 7–10 months of age were vaccinated while an equal number of untreated children served as controls. Since the majority of children in both groups developed somatitis, the vaccine was deemed a failure. The first conduct of a prospective placebo-controlled trial may have been that reported by KERN and SCHIFF, in which inactivated whole virus vaccine was tested for therapeutic efficacy (KERN and SCHIFF 1964). While 70% of the vaccine recipients reported improvement, 76% of the placebo recipients also reported improvement. Thus, this vaccine trial was a dramatic example of the importance of conducting an appropriately designed and controlled study.

3.1.1 Lupidon Vaccine

In Europe, heat-killed whole virus vaccines, Lupidon H from HSV-1 and Lupidon G from HSV-2 (prepared by the Hermal-Chemie Company, Germany), have been widely available (NASEMANN 1976). A large scale, open trial of this preparation for therapeutic treatment of recurrent genital herpes resulted in clinical improvement in 81% of 1059 participants (SCHMERSAHL and RUDIGER 1975). Injections were administered at closely spaced 1–2 week intervals until clinical improvement was noted. Maintenance injections were then given every 3–6 months for years. No placebo-controlled trials have been reported other than anecdotally.

3.1.2 Dundarov Vaccine

In Bulgaria, a formalin-inactivated polyvalent vaccine consisting of preparations of multiple strains of HSV-1, HSV-2, or both, grown in rabbit embryo kidney cells, has been licensed since 1975 (DUNDAROV et al. 1982). A total of 2350 patients with recurrent HSV-1 or HSV-2 disease were treated in open trials conducted by many different physicians with a series of eight closely spaced initial immunizations at 2–3 day intervals followed by six injections at 20 day intervals. Additional injections were administered in the absence or the waning of a therapeutic effect. In 60%–65% of the patients, the titer of HSV antibody and the magnitude of the HSV cellular immune response increased but there were no correlations between immune response and clinical response. Within 6 months after the initiation of therapy, 85% of the patients reported improvement and 65% reported significant improvement. With longer follow-up periods of 4–5 years, these indices increased to 95% and 80%.

3.1.3 Skinner Vaccine

Many workers have attempted to improve the safety of HSV whole virus vaccines by removing the viral DNA. One of the most widely reported preparations of this type is the "antigenoid" HSV vaccine developed by SKINNER et al. (1982a). This preparation has been tested as both a therapeutic and as a prophylactic vaccine with consistently encouraging claims of efficacy published for both applications (SKINNER et al. 1982a, 1989). The SKINNER vaccine is a formalin-treated detergent (NP40) extract of HSV-1-infected human embryo lung cells (MRC 5). The vaccine contains virtually all of the polypeptides synthesized by virus-infected cells based on comparisons of SDS-PAGE profiles (SKINNER et al. 1982b). gD is present at a concentration of at least 0.6 μg dose (0.2%) prepared from 2×10^7 virus-infected cells (SKINNER et al. 1987). There is <50 pg viral DNA per dose (SKINNER et al. 1987). Several prophylactic trials have been conducted in couples in which only one partner has genital herpes and the second partner is at risk of acquisition (SKINNER et al. 1987, 1989). In one such study, 50 consorts of patients with a history of genital HSV received one, two, or three vaccinations at an unspecified interval; 44 of these subjects received the vaccine with 50% alum (Alhydrogel) as adjuvant using an unspecified dose (SKINNER et al. 1987). Only one of the 50 participants developed symptoms of genital herpes during a 4–48 month follow-up period, corresponding to an annual acquisition rate of 1.7%. However, there was no reported evaluation of the HSV serological status of the participants prior to or after immunization and no evaluation of the immunogenicity of the vaccine was included. The claim of efficacy was based on the rate of disease contraction compared with an undocumented 25% rate in unvaccinated, historical controls. In another study, only one of 60 vaccinated consorts acquired HSV during an average 18 months follow-up (SKINNER et al. 1982b); a third study of 15 consorts revealed no HSV acquisition over a 24–36 month follow-up period (MUNIU et al. 1987); and a fourth study revealed that only one of 101 subjects at risk of HSV (some of whom were included in an earlier report; (SKINNER et al. 1987) was infected during an average 20.6 month follow-up period (SKINNER et al. 1989). All of these trials were conducted in an open fashion with results compared to historical controls; thus, it is impossible to determine the merit of the claims of efficacy. Also, little data on the immune responses to vaccine were reported, except that HSV neutralizing antibody was detected in 59% of low dose (2×10^7 infected cell equivalents) and 90% of high dose (1×10^8 infected cell equivalents) recipients (SKINNER et al. 1982b). Therapeutic trials revealed that patients with one clinical HSV episode prior to immunization had fewer recurrences than retrospective controls (SKINNER et al. 1982b; WOODMAN et al. 1983) and patients with established recurrent genital HSV reported improvement (SKINNER et al. 1982b). When vaccine was administered after acute infection, the recurrence rate decreased to 31% compared to 85% for unvaccinated controls (WOODMAN et al. 1983). These trials were also conducted in an open format with no placebo control, although the authors noted the necessity to "subject these data to the scrutiny of a placebo-controlled trial with an objective system of clinical assessment" (SKINNER et al. 1982b).

3.1.4 Cappel Vaccine

CAPPEL and coworkers prepared a DNA-free HSV-2 subunit vaccine by detergent (NP40) disruption of HSV-2 purified after growth in chicken embryo fibroblasts. The virion envelope proteins were separated from the nucleocapsid by ultracentrifugation on a sucrose gradient and adsorbed on a 2% solution of $Al(OH)_2$ (CAPPEL et al. 1985). The vaccine contains at least nine polypeptides including the three major glycoproteins. A total of 77 individuals were immunized three times at a 2 week interval with 1.5 μg/kg body weight of vaccine (CAPPEL et al.

1985). Of these, 18 had no past evidence of HSV infection, 15 had severe recurrent HSV-1 oral recurrences, and 44 had recurrent HSV-2 infection; an additional 18 with recurrent genital HSV received no vaccine and served as a control group. HSV-specific antibody responses were measured by four different assays; neutralization, immunofluorescence, complement fixation, and complement-dependent cytotoxicity. The antibody responses of patients with recurrent genital HSV-2 were boosted by immunization with increases of 1.5-, 1.6-, 1.7-, and 6.4-fold, respectively, for each of these indices when measured 2 weeks after the third immunization. For those individuals who were initially HSV-seronegative, the vaccine elicited immune responses that were similar to those generated by HSV-2 infection, being 1.1-, 0.5-, 0.6-, and 4.1-fold, respectively, those reported for the HSV-2-seropositive individuals at study entry. The vaccine induced cellular immune responses to HSV antigen as detected by lymphoproliferation in the 18 seronegatives, and this response increased in the 44 seropositives following vaccination. The vaccine appeared to have therapeutic efficacy reducing both the number of recurrences and the time to healing. Similar data were reported with a smaller number of patients, 15 seronegatives and 28 with recurrent HSV, in an earlier study (CAPPEL et al. 1982). These results, particularly the immunological responses, are certainly encouraging, but given the large placebo effect observed in past trials (KERN and SCHIFF 1964; KUTINOVÁ et al. 1988) the claims of efficacy would have to be confirmed by double-blind placebo-controlled trials.

3.1.5 Kutinova Vaccine

KUTINOVA and coworkers conducted a double-blind placebo-controlled trial in Czechoslovakia of an HSV-1-derived glycoprotein vaccine for immunotherapeutic treatment of patients with frequent recurrent HSV-1 or HSV-2 infections (KUTINOVÁ et al. 1988). Their earlier studies with similar preparations had demonstrated that the vaccine induced neutralizing antibody responses in several animal species and conferred protection against an intraperitoneal challenge infection and the establishment of latency with HSV-1 or HSV-2 in mice (KUTINOVÁ et al. 1982). The vaccine was prepared from a detergent extract (Triton X-100) of HSV-1-infected human embryo lung cells which was treated with DNase I, chromatographed on a concanavalin A lectin column, treated with formalin, and adsorbed to aluminum hydroxide. From the description, this preparation may be similar to that manufactured by Merck, Sharp and Dohme (LARSON and LEHMAN 1983) and tested by MERTZ et al. (1988) as described below. A total of 42 individuals with culture-confirmed recurrent herpes, 18 with HSV-1, 24 with HSV-2, and all experiencing at least four episodes per annum, were enrolled. Matched patients were immunized three times at 3 week intervals with vaccine or placebo and then followed for 14 months. The vaccine was not effective since 42.8% of vaccine recipients experienced a twofold or greater decrease in the frequency of recrudescences compared to 35.0% of those who received placebo. Moreover, the vaccine was poorly immunogenic, increasing neutralizing antibody titers in only 1 of 9 HSV-1$^+$ and 2 of 12 HSV-2$^+$ individuals. It is interesting that, although the vaccine was not effective, the process of vaccination was; again a very large placebo effect was observed.

3.1.6 Summary of Earlier Human HSV Trials

Broadly scanning the human experience with various HSV vaccine candidates prior to 1990, there are several notable points: (1) There has been a long and durable interest in the development of such vaccines, with trials conducted as early as the 1920s. The majority of these trials have been directed towards a therapeutic application; few trials have been conducted to test prophylactic efficacy. (2) Many diverse approaches have been employed, including even autoinoculation with the patient's own live virus. The predominant approach has been the use of preparations of infected cell extract or partially purified infected cell extract. (3) With rare exceptions, trials have been conducted in an open format with encouraging claims of efficacy made based on the use of historical controls or of patients' recollections of pretreatment disease patterns. When appropriate placebo-controlled trials have been conducted, no efficacious vaccines have emerged. In general, the vaccine preparations were poorly described and characterized. Likewise, evaluations of the immune response to vaccination were often lacking or scantly described. In the absence of a successful vaccine, it is difficult to determine why these previous candidates were not useful. In general, it may be that infected cell extracts contain low concentrations of viral proteins. In support of this hypothesis, the immunogenicity (when reported) of many preparations was modest. In addition, many vaccines utilized HSV-1-derived proteins to protect against or treat HSV-2 infections. Many animal experiments now demonstrate the increased level of protection achieved with a homotypic versus a heterotypic preparation (BURKE et al. 1989a). It is also of interest that, whereas past therapeutic vaccine trials revealed a very large placebo effect, 75% in the study of KERN and SCHIFF (1964) and 35% in the study of KUTINOVA et al. (1988), therapeutic trials of acyclovir have not revealed a significant placebo effect (STRAUS et al. 1984; DOUGLAS et al. 1984). This difference is most likely related to differences in trial design and clinical endpoints.

3.2 Recent Human HSV Vaccine Experience

3.2.1 Merk Vaccine

There is a continued interest in the development of HSV vaccines as evidenced by the recent initiation of several different human clinical trials. The first of these was conducted by COREY and colleagues at the University of Washington,

Seattle. As reported by MERTZ et al. (1988), a subunit vaccine composed of a mixture of viral glycoproteins, prepared by MERCK, SHARP, and DOHME, was evaluated for prophylactic efficacy in a double-blind placebo-controlled trial. The vaccine was prepared by extracting HSV-2-infected chicken embryo fibroblasts with a buffer containing 2% Triton X-100 and 4.5% ethanol followed by centrifugation and DNase treatment of the clarified supernatant. A glycoprotein enriched fraction was isolated by chromatography on a lentil lectin affinity column and elution with α-methyl mannoside. The mixture was then treated with formalin and adsorbed to alum (LARSON et al. 1983). Although, the exact composition of the mixture was not determined, both host cell and viral-encoded glycoproteins were isolated by this procedure and select HSV glycoproteins, including gB, gC, gD, gE, and gG, were confirmed to be present (LARSON et al. 1983; ASHLEY et al. 1985b). The vaccine was safe and well tolerated and elicited an antibody response to gB, gD, and other glycoproteins detectable by ELISA and western blot assay. This antibody was capable of neutralizing virus in vitro and supporting antibody-dependent cellular cytotoxicity (MERTZ et al. 1984; ASHLEY et al. 1985). HSV-seronegative vaccine recipients developed a lymphoproliferative response following vaccination and, in initially HSV-seropositive vaccinees, the magnitude of this response was increased by immunization (MERTZ et al. 1984; ZARLING et al. 1988). Cytolytic T cell clones were readily isolated from vaccinees (ZARLING et al. 1988); however, these clones were all HLA class II-restricted and thus their potential role in the control or prevention of HSV infection is not clear. A total of 161 HSV discordant couples, in which one individual had no serological or clinical evidence of genital HSV and was at risk of acquiring the infection from their sexual partner, participated in the efficacy trial. Volunteers were immunized three times at months 0, 1, and 6 with 50 μg of either glycoprotein mixture or placebo bound to alum. Antibody responses were monitored by neutralization, western blot, and ELISA. Acquisition of symptomatic HSV disease was confirmed by clinical observation and by viral culture. Asymptomatic infection was detected by the development of antibody responses to abundant virion proteins not present in the glycoprotein mixture such as the major capsid protein VP5 or the virion structural proteins, the ICP35 family (ASHLEY et al. 1987). During the course of the trial, 14 individuals acquired an HSV infection, 9 of the vaccine and 5 of the placebo recipients, corresponding to annual rates of infection of 11.3% and 7.9%, respectively. Thus, this vaccine failed to provide protection from acquisition of genital HSV infection. Based on the sample size and the trial design, the study had a 50% chance of detecting a 50% decrease in the acquisition rate. The lack of efficacy may have been related in part to the low immunogenicity of the vaccine. In initially seronegative vaccinees, peak antibody responses to gB and gD, measured 1 month following the third immunization by ELISA, were less than those elicited by natural infection with HSV-2. In addition, the antibody response was not durable, since at 12 months gD2 antibody was no longer detectable in 3 of 15 subjects who had developed antibody to gD2. In initially HSV-1-seropositive vaccinees, antibody titers to gB2 and gD2 increased following each vaccine dose to a peak geometric mean titer about two-fold greater than that present at study entry. Finally, there was no apparent difference in the immune response to vaccine for matched samples of those who did and who did not become infected. Despite the lack of efficacy, the trial serves as an example of a useful design for the conduct of a blind and controlled efficacy trial. In addition, the trial helped to identify risk factors associated with the transmission and acquisition of genital herpes (MERTZ et al. 1991). For example, prior HSV-1 infection reduced the risk of acquisition of HSV-2 infection by 3.5-fold, from an annual rate of 31.8% in doubly seronegative females to 9.1% in those with prior HSV-1 antibody at study entry. Transmission occurred more often from a male (16.9%) than from a female (3.8%) source partner ($p = 0.05$). From histories collected near the time of transmission, 9 of 13 events occurred asymptomatically and, in the remainder, occurred within prodrome or before the development of recognizable lesions. This data on transmission rates and relative risk factors is helpful for the design of future clinical trials and underscores the importance of the stratification of participants based on their initial HSV serological status.

3.2.2 Skinner Vaccine

A double-blind placebo-controlled trial of the Skinner vaccine, prepared by Porton International, has been conducted at Rush Memorial Medical Center in Chicago, IL by Benson and colleagues (C. BENSON, personal communication). This study was designed to assess therapeutic efficacy in individuals with culture-documented genital HSV disease with at least six recurrences per year or ≧30 days of genital lesions for 6 months. Participants received 0.4 ml (300 μg) of the Skinner vaccine ($n = 134$) or placebo ($n = 123$) adsorbed to alum and administered subcutaneously on days 0, 30, and 60. They were monitored for HSV disease for 14 months. A clinical response was defined as greater than a 50% decrease in recurrent disease frequency compared to the pretreatment frequency. The collation and interpretation of the data are still in progress. However, a preliminary analysis suggests that there was no difference in the mean time to first recurrence or in the mean number of outbreaks per patient between the vaccine and placebo recipients. Although the number of clinical responders in the vaccinated group was greater than in the placebo group, this difference was not statistically significant ($p = 0.054$). The vaccine boosted HSV-specific lymphoproliferative and antibody responses. However, it remains to be determined if the magnitude of the immunological response correlated in any way with the clinical disease response. The completed analysis of this carefully conducted trial should provide valuable information to guide the design and testing of future immunotherapeutic vaccines. The study represents the first blind and controlled trial of the Skinner vaccine following greater than 10 years of extensive testing in open trials in the United Kingdom, which resulted in repeated claims of efficacy (discussed previously). The potential prophylactic efficacy of the Skinner vaccine now remains to be tested in a similar fashion in a blind and placebo-controlled trial.

3.2.3 Lederle Vaccine

Recently a phase I safety and immunogenicity trial of a subunit vaccine composed of gD1, purified by immunoaffinity chromatography from HSV-1-infected Vero cells, was conducted by FRENKEL et al. working with Bryson at the University of California, Los Angeles and prepared by Lederle Biologicals (FRENKEL et al. 1990). There were 24 males who were HSV-1-seropositive by western blot analysis and who received a single dose of placebo or 2.5, 5, or 10 μg of gD1 protein bound to 500 μg of alum. The vaccine was well tolerated with minimal local and systemic reactions. Immune responses were monitored by western blot, in vitro neutralization, and lymphoproliferation assay. Complete analyses of the immune responses are still in progress. However, no increase in the antibody titer to gD was detected by western blot assay and only minimal changes in the neutralization titer were observed. An increase in the mean lymphoproliferative response, measured as the stimulation index relative to the value at study entry, was detected for the 10 μg dose group when measured at days 2 and 7 postvaccination but this increase was no longer apparent at weeks 4 and 8. No change was noted for the lower dose groups. The apparently modest immunogenicity of this vaccine preparation is somewhat surprising since a similar vaccine has proven to be far more immunogenic, as described below.

3.2.4 Biocine Vaccine

In a second HSV phase I trial also recently initiated, a different preparation of gD was tested for safety and immunogenicity by FREIFELD et al. (1990) at the National Institutes of Health (NIH). gD2 was produced by recombinant technology as a truncated derivative lacking the transmembrane and cytoplasmic domains. It was secreted from Chinese hamster ovary cells and purified by a series of conventional chromatography steps. The protein is glycosylated and folded like the native protein prepared from virus-infected cells. Two doses, consisting of 30 and 100 μg of protein bound to 400 μg of alum, were administered to a total of 24 individuals of three different initial serological status groups, 8 who were doubly seronegative, 8 with solely HSV-1 antibody, and 8 who had HSV-2 antibody with and without HSV-1 antibody as determined by western blot analysis (FREIFELD et al. 1990). Participants were vaccinated four times, receiving an initial series of three immunizations at 0, 1, and 2 months and a booster dose at 12 months (FREIFELD et al. 1990; PARR et al. 1991). The vaccine was well tolerated and caused only minor local discomfort. Antibody response to gD2 was determined by ELISA assay. The baseline geometric mean titers (GMTs) were <27 for the HSV $1^-/2^-$ subjects, 684 for the HSV $1^+/2^-$ subjects, and 1845 for the HSV $1^\pm/2^+$ individuals. For the seronegative recipients of the 100 μg dose, the GMT peaked at 426 at month 3, 1 month after the first three immunizations, fell by 57% to 186 by 12 months, and was boosted to 4552 at 13 months. Seronegative subjects that received the 30 μg dose had GMTs two- to three-fold lower than those of the 100 μg dose group. The response in initially seropositive individuals was similar for both dose groups, with GMTs at 3 months about 15-fold above the baseline values, reaching 28919 for recipients of 100 μg. This antibody response was durable, falling only 20% by month 12, and was restored to the previous high value by the booster immunization. HSV-2 neutralizing antibody titers paralleled the ELISA titers with five- to seven-fold average increases in the GMT compared to the initial value for the seropositive individuals. The baseline lymphoproliferative response to soluble gD2 antigen in the seropositive individuals was so high that it was not possible to detect any increases associated with vaccination. It is possible that the use of a quantitative assay, such as limiting dilution, rather than a standard qualitative assay could reveal potential changes in the frequency of HSV-specific T cells resulting from vaccination. This trial is still in progress, with patients to be followed for a total of 24 months.

As determined by ELISA titer, the peak antibody response in HSV-$1^-/2^-$ individuals receiving the 100 μg dose exceeded that elicited by natural HSV-2 infection by 2.5-fold, while those developed by the HSV $1^+/2^-$ group increased six-fold above their initial value. The elicitation of these high antibody titers is encouraging but their biological relevance remains to be tested in appropriately controlled and blind studies. A placebo-controlled double-blind therapeutic study of this gD2/alum vaccine has been initiated in individuals with frequent recurrent genital HSV. It should be noted that, when similar gD2/alum vaccines were tested for immunotherapeutic efficacy in a guinea pig model of genital herpes, they failed to reduce recurrence rates in contrast to other vaccine formulations (BERMAN et al. 1988; BURKE and HO, unpublished results). However, the guinea pig model mimics acute intervention immediately following primary infection. Moreover, the magnitude and durability of the antibody response in humans were not predicted by the results in guinea pigs. These clinical trials will reveal, in part, if the guinea pig animal model has any predictive value for the future development of therapeutic HSV vaccines. Additional studies testing the safety and immunogenicity of a mixture of two recombinant glycoproteins, gB2 and gD2, combined with the muramyl dipeptide derivative MTP-PE as adjuvant in a low oil emulsion are planned.

The difference in the antibody responses observed in these two recent phase I trials evaluating the immunogenicity of gD bound to alum is puzzling. It should be noted that lower doses of gD1, i.e., 2.5, 5, and 10 μg, were tested in the first trial compared to the 30 and 100 μg doses of gD2 used in the second trial. Other differences that may account for the results include the adjuvant potency of the specific alum preparations and the particular assays employed to evaluate the immune responses.

4 Conclusions

The development of an effective HSV vaccine is essential to limit the continuing spread of HSV infection. Although there is a long and disappointing history of herpes vaccine trials, recent advances in the production of both live virus and subunit vaccine candidates appear promising. However, there are currently severe limitations in our ability to approach the rational rather than the pragmatic design of new vaccine candidates: (1) A better understanding of the mechanisms of protection engendered by natural infection and, in particular, the role of cellular and mucosal immunity is required. (2) In vitro measurable correlates of in vivo protection need to be clearly identified. (3) The validity and utility of various animals models for predicting human immune response need to be better established. (4) An understanding of the immune responses that control or limit recurrent disease in humans would facilitate the development of immunotherapeutic vaccines. Despite these limitations, the next decade of research may bring rapid advances as our knowledge of the importance of particular HSV-specific immune responses increases in concert with an evaluation of the immunogenicity and efficacy of the current vaccine candidates.

Acknowledgments. I would like to recognize the invaluable contributions of my many colleagues at Chiron Corporation, including Gary Ott and Gary Van Nest for their studies on adjuvant formulations; Peter Brown, Barbara Gervase, Celine Hu, and Frank Marcus for large scale protein production and purification; Cheryl Goldbeck and Philip Ng for animal studies and immunoassays; Corry Dekker and Sue Adair for organization and conduct of human clinical trials; Dino Dina for his continuous support; and our collaborators, Larry Stanberry, Rodney Ho, and Steve Straus, for their studies in animals and humans. I also thank Jeffrey Neidleman for help with the preparation of this manuscript.

References

Adams SE, Dawson KM, Gull K, Kingsman SM, Kingsman AJ (1987) The expression of hybrid HIV: Ty virus-like particles in yeast. Nature 329: 68

Anderson FD, Ushijima RN, Larson CL (1974) Recurrent herpes genitalis: treatment with *Mycobacterium bovis* (BCG). Obstet Gynaecol 43: 797

Anderson SG, Hamilton J, Williams S (1950) An attempt to vaccinate against herpes simplex. Aust J Exp Biol Med Sci 28: 579

Arbeter AM, Starr SE, Weibel RE, Plotkin SA (1986) Varicella vaccine studies in healthy children and adults. Pediatrics 78: 748–756

Ashley R, Benedetti J, Corey L (1985a) Humoral immune response to HSV-1 and HSV-2 viral proteins in patients with primary genital herpes. J Med Virol 17: 153–166

Ashley R, Mertz G, Clark H, Schick M, Salter D, Corey L (1985b) Humoral immune response to herpes simplex virus type 2 glycoproteins in patients receiving a glycoprotein subunit vaccine. J Virol 56: 475–481

Ashley R, Mertz GJ, Corey L (1987) Detection of asymptomatic herpes simplex virus infections after vaccination. J Virol 61: 264–268

Berman PW, Vogt PE, Gregory T, Lasky LA, Kern ER (1988) Efficacy of recombinant glycoprotein D subunit vaccines on the development of primary, recurrent, and latent genital infections with herpes simplex type 2 in guinea pigs. J Infect Dis 157(5): 897–902

Bierman SM (1976) BCG immunoprophylaxis of recurrent herpes genitalis. Arch Dermatol 112: 1410

Blacklaws BA, Krishna S, Minson AC, Nash AA (1990) Immunogenicity of herpes simplex virus type I glycoproteins expressed in vaccinia virus recombinants. Virology 177: 727–736

Bonneau RH, Jennings SR (1988) Modulation of acute and latent herpes simplex virus infection in C57BL/6 mice by adoptive transfer of immune lymphocytes with cytolytic activity. J Virol 63: 1480–1484

Brain RT (1936) Biological therapy in virus diseases. Br J Dermatol Syph 48: 21

Brynestad K, Babbit B, Huang L, Rouse BT (1990) Influence of peptide acylation, liposome incorporation, and synthetic immunomodulators on the immunogenicity of a 1–23 peptide of glycoprotein D of herpes simplex virus: implications for subunit vaccines. J Virol 64: 680–685
Burke KL, Dunn G, Ferguson M, Minor PD, Almond JW (1988) Antigen chimeras of poliovirus. Nature 332: 81
Burke RL (1991) Development of a herpes simplex virus subunit vaccine for prophylactic and therapeutic use. Rev Infect Dis 13 (Supplement 11): 5906–5911
Burke RL, Van Nest G, Carlson J et al. (1989a) Development of a herpes simplex virus subunit vaccine. Vaccines 89: 337–382
Burke RL, Van Nest G, Gervase B et al. (1989b) Development of an HSV subunit vaccine: effect of adjuvant composition and antigen dose. In: Meheus A, Spier RE (eds) Vaccines for sexually transmitted diseases. Butterworth, Stoneham, pp 191–196
Burke RL, Hartog K, Croen KD, Ostrove JM (1991) Detection and characterization of latent HSV RNA by in situ and northern blot hybridization in guinea pigs. Virology 181: 793–797
Cantin EM, Eberle R, Baldick JL, Moss B, Willey DE, Notkins AL, Openshaw H (1983) Expression of herpes simplex virus 1 glycoprotein B by a recombinant vaccinia virus and protection of mice against lethal HSV-1 challenge. Proc Natl Acad Sci USA 80: 7155–7159
Cappel R, Sprecher S, de Cuyper F (1982) Immune responses to DNA free herpes simplex proteins in man. Dev Biol Stand 52: 345–350
Cappel R, Sprecher S, DeCuyper F, DeBraekeleer J (1985) Clinical efficacy of a herpes simplex subunit vaccine. J Med Virol 16: 137–145
Clements JD (1989) Use of attenuated mutants of *Salmonella* as vaccine vectors. Vaccines 89: 283–290
Cremer KJ, Mackett M, Wohlenberg C, Notkins AL, Moss B (1985) Vaccinia virus recombinant expressing herpes simplex virus type 1 glycoprotein D prevents latent herpes in mice. Science 228: 737–740
Dix RD (1987) Prospects for a vaccine against herpes simplex virus types 1 and 2. Prog Med Virol 34: 89–128
Dix RD, Mills J (1985) Acute and latent herpes simplex virus neurological disease in mice immunized with purified virus-specific glycoproteins gB or gD. J Med Virol 1785: 9–18
Douglas JM, Critchlow C, Benedetti J, Mertz GJ, Connor JD, Hintz MA, Fahnlander A, Remington M, Winter C, Corey L (1984) A double-blind study of oral acyclovir for suppression of recurrences of genital herpes simplex virus infection. N Engl J Med 310: 1551–1556
Douglas JM, Vontver A, Stamm WM, Reeves WC, Critchlow C, Remington ML, Holmes KK, Corey L (1985) Ineffectiveness and toxicity of BCG vaccine for the prevention of recurrent genital herpes. Antimicrob Agents Chemother 27: 203
Dundarov S, Andonov P, Bakalov B, Nechev K, Tomov C (1982) Immunotherapy with inactivated polyvalent herpes vaccines. Dev Biol Stand 52: 351–358
Eberle R, Mou S-W, Zaia JA (1985) The immune response to herpes simplex virus: comparison of the specificity and relative titers of serum antibodies directed against viral polypeptides following primary herpes simplex virus type 1 infections. J Med Virol 16: 147–162
Eisenberg RJ, Cerini CP, Heilman CJ, Joseph AD, Dietzschold B, Golub E, Long D, Ponce de Leon M, Cohen GH (1985) Synthetic glycoprotein D-related peptides protect mice against herpes simplex virus challenge. J Virol 56: 1014–1017
Erturk M, Jennings R, Hockley D, Potter CW (1989) Antibody responses and protection in mice immunized with herpes simplex virus type 1 antigen immune-stimulating complex preparations. J Gen Virol 70: 2149–2156
Forrester AJ, Sullivan V, Simmons A, Blacklaws BA, Smith GL, Nash AA, Minson AC (1991) Induction of protective immunity with antibody to herpes simplex virus type I glycoprotein H (gH) and analysis of the immune response to gH expressed in recombinant vaccinia virus. J Gen Virol 72: 369–375
Foster PD, Abshier AB (1937) Smallpox vaccine in treatment of recurrent herpes. Arch Dermatol Syphl 36: 294
Frank SB (1938) Formolised herpes virus therapy and the neutralising substance in herpes simplex. J Invest Dermatol 1: 267
Freifeld AG, Savarese B, Krause PR, Paar D, Adair S, Tigges MA, Burke RL, Straus SE (1990) Phase I testing of a recombinant herpes simplex type 2 glycoprotein D vaccine. Interscience conference on antimicrobials and chemotherapeutics 722 (abstract)
Frenchick PJ, Sabara MIJ, Babiuk LA (1989) Use of a viral nucleocapsid particle as a carrier for synthetic peptides. Vaccines 89: 479–484

Frenkel LM, Dillon M, Garratty E, Bryson Y, Hackell J, Cawein A, Abramovitz A, Miskin E (1990) A randomized double blind, placebo-controlled phase I, trial of a herpes simplex virus purified glycoprotein (gDI) vaccine. Interscience conference on antimicrobials and chemotherapeutics 721 (abstract)
Freund H (1928) Beiträge zue Kenntnis von Herpes und Zoster. Arch Dermatol Syphilol 155: 282
Fuller AO, Santos RE, Spear PG (1989) Neutralizing antibodies specific for glycoprotein H of herpes simplex virus permit viral attachment to cells but prevent penetration. J Virol 63: 3435–3443
Geerlings HJ, Weijer WJ, Welling GW, Welling-Wester S (1989) The influence of different adjuvants on the immune response to a synthetic peptide comprising amino acid residues 9–21 of herpes simplex virus type 1 glycoprotein D. J. Immunol Methods 124: 95–102
Gompels U, Minson A (1986) The properties and sequence of glycoprotein H of herpes simplex virus type 1. Virology 153: 230–247
Hall MJ, Katrak K (1986) The quest for a herpes simplex virus vaccine: background and recent developments. Vaccine 4: 138–149
Heber-Katz E, Valentine S, Dietzschold B, Burns-Purzycki C (1988) Overlapping T cell antigenic sites on a synthetic peptide fragment from herpes simplex virus glycoprotein D, the degenerate MHC restriction elicited, and functional evidence for antigen-Ia interaction. J Exp Med 167: 275–287
Hilfenhaus J, Christ H, Kohler R, Moser H, Kirchner H, Levy HB (1981) Protectivity of herpes simplex virus antigens: studies in mice on the adjuvant effect of PICLC and on the dependence of protection on T cell competence. Med Microbiol Immunol 169: 225–235
Ho RJY, Burke RL, Merigan TC (1989) Antigen presenting liposomes are effective in treatment of recurrent herpes simplex genitalis in guinea pigs. J Virol 63: 2951–2958
Ho RJY, Burke RL, Merigan TC (1990) Physical and biological characterization of antigen presenting liposome formulations: relative efficacy for the treatment of recurrent genital HSV-2 in guinea pigs. Antiviral Res 13: 187–200
Javier RT, Sedarati F, Stevens JG (1986) Two avirulent herpes simplex viruses generate lethal recombinants in vivo. Science 234: 746–748
Johnson RE, Nahmias AJ, Magder LS, Lee FK, Brooks CA, Snowden CB (1989) A seroepidemiologic survey of the prevalence of herpes simplex virus type 2 Infection in the United States. N Eng J Med 321: 7–12
Kern AB, Schiff BL (1964) Vaccine therapy in recurrent herpes simplex. Dermatologica 89: 844–845
Kino Y, Nozaki C, Nakatake H, Mizuno K, Mori R (1989) Immunogenicity of herpes simplex virus glycoprotein gB-1-related protein produced in yeast. Vaccine 7: 155–160
Kit S (1982) Problems and progress in the development of vaccines against genital herpes. Infect Dis 12: 4–9
Klein RJ, Buimovici-Klein E, Moser H, Moucha R, Hilfenhaus J (1981) Efficacy of a virion envelope herpes simplex virus vaccine against experimental skin infections in hairless mice. Arch Virol 68: 73–80
Krishna S, Blacklaws BA, Overton HA, Bishop DHL, Nash AA (1989) Expression of glycoprotein D of herpes simplex virus type 1 in a recombinant baculovirus protective responses and T cell recognition of the recombinant-infected cell extracts. J Gen Virol 70: 1805–1814
Kutinová L, Vonka V, Šlicktová V (1980) Immunogenicity of subviral herpes simplex virus preparations: protection of mice against intraperitoneal infection with live virus. Acta Virol 24: 391–398
Kutinová L, Šlichtová V, Vonka V (1982) Subvlral herpes simplex vaccine. Dev Biol Stand 52: 313–319
Kutinová L, Benda R, Kaloš Z, Dhalý V, Votruba T, Kvičalová E, Petroská P, Doutllk S, Kamlnková J, Demorázková E, Bergmanová V, Krčmár M, Suchánková A, Němečková S, Vonka V (1988) Placebo-controlled study with subunit herpes simplex virus vaccine in subjects suffering from frequent herpetic recurrences. Vaccine 6: 223–228
Larson VM, Lehman ED (inventors) (1983)Herpes subunit vaccine. US patent 4,374,127. Merck & Co. Inc, assignee. Feb 15, 16 p. Int CL^3A61K 39/245
Lebkowski JS, McNally MM, Okarma TB, Lerch LB (1988) Adeno-associated virus: a vector system for efficient introduction and integration of DNA into a variety of mammalian cell types. Mol Cell Biol 8: 3988–3996
Martin S, Rouse BT (1987) The mechanisms of antiviral immunity induced by a vaccinia virus recombinant expressing herpes simplex virus type 1 glycoprotein D: clearance of local infection. J Immunol 138: 3431–3437
Martin S, Moss B, Berman PW, Laskey LA, Rouse BT (1987) Mechanisms of antiviral immunity induced by a vaccinia virus recombinant expressing herpes simplex virus type 1 glycoprotein D: cytotoxic T cells. J Virol 61: 726

Martin S, Cantin E, Rouse BT (1989) Evaluation of antiviral immunity using vaccinia virus recombinants expressing cloned genes for herpes simplex virus type I glycoproteins. J Gen Virol 70: 1359–1370
Martin S, Zhu XX, Silverstein SJ, Courtney RJ, Yao F, Jenkins FJ, Rouse BT (1990) Murine cytotoxic T lymphocytes specific for herpes simplex virus type 1 recognize the immediate early protein ICP4 but not ICPO. J Gen Virol 71: 2391–2399
McDermott MR, Smiley JR, Brais PLJ, Rutzroga HE, Brenenstock J (1985) Immunity in the female genital tract after intravaginal vaccination of mice with an attenuated strain of herpes simplex virus type 2. J Virol 51: 747–753
McDermott MR, Graham FL, Hanke T, Johnson DC (1989) Protection of mice against lethal challenge with herpes simplex virus by vaccination with an adenovirus vector expressing HSV glycoprotein B. Virology 169: 244–247
McLaughlin-Taylor E, Wiley DE, Cantin EM, Eberle R, Moss B, Openshaw H (1988) A recombinant vaccinia virus expressing herpes simplex virus type 1 glycoprotein B induces cytotoxic T lymphocytes in mice. J Gen Virol 69: 1731
Meignier B (1985) Vaccination against herpes simplex virus infections. In: Roizman B, Lopez C (eds) Herpesviruses, vol 4. Plenum, New York, pp 265–296
Meignier B, Jourdier TM, Norrild B, Pereira L, Roizman B (1987a) Immunization of experimental animals with reconstituted glycoprotein mixtures of herpes simplex virus 1 and 2: protection against challenge with virulent virus. J Infect Dis 155: 921–931
Meignier B, Longnecker R, Roizman B (1987b) Construction and in vivo evaluation of two genetically engineered prototypes of live attenuated herpes simplex virus vaccines. Vaccines 87: 368–373
Meignier B, Longnecker R, Roizman B (1988) In vivo behavior of genetically engineered herpes simplex viruses R7017 and R7020: construction and evaluation in rodents. J Infect Dis 158: 602–614
Meignier B, Martin B, Whitley RJ, Roizman B (1990) In vivo behavior of genetically engineered herpes simplex viruses R7017 and R7020. II. Studies in immunocompetent and immunosuppressed owl monkeys (Aotus trivirgatus). J Infect Dis 162: 313–321
Mertz GJ, Peterman G, Ashley R, Jourden JL, Salter D, Morrison L, McLean A, Corey L (1984) Herpes simplex virus type 2 glycoprotein-subunit vaccine: tolerance and humoral and cellular responses in humans. J Infect Dis 150: 242–249
Mertz GJ, Jones CC, Mills J, Fife KH, Lehrman SN, Stapleton JT, Hill EL, Davis G, and the Acyclovir Study Group (1988) Long-term acyclovir suppression of frequently recurring genital herpes simplex virus infection. A multicenter double-blind trial. JAMA 260: 201–206
Mertz GJ, Benedetti J, Ashley R, Selke S, Corey L (1992) Risk factors for the sexual transmission of genital herpes. Ann Int Med (in press)
Muggeridge MI, Roberts SR, Isola VJ et al. (1990) Herpes simplex virus. In: Van Regenmortel MHV, Neuraht AR (eds) Immunochemistry of viruses, II. The basis for serodiagnosis and vaccines. Elsevier, New York, pp 459–481
Muniu EM, Durham J, Shariff D, Hartley CE, Fuller A, Melling J, Wiblin C, Wilkins G, Buchan A, Skinner GRB (1987) Antibody reactivity with Skinner HSV vaccine. Med Microbiol Immunol (Berl) 176: 315–327
Nasemann T (1976) Herpes simplex type II vaccine: the favorable European experience. Int J Dermatol 15: 587–588
Paoletti E, Lipinskas BP, Samsonoff C, Mercer S, Panicali D (1984) Construction of live vaccines using genetically engineered poviruses: biological activity of vaccinia virus recombinants expressing the hepatitus B virus surface antigen and the herpes simplex virus glycoprotein D. Proc Natl Acad Sci USA 81: 193–197
Parr D, Savarese B, Burke RL, Margolis D, Meier J, Markoff L, Ashley R, Corey L, Adair S, Dekker C, Straus SE (1991) Ability of a recombinant herpes simplex virus type 2 glycoprotein D vaccine to induce antibody titers comparable to those following genital herpes. Clin Res 39: 216A
Sanchez-Pescador L, Burke RL, Ott G et al. (1988a) A comparison of adjuvant efficacy for a recombinant herpes simplex glycoprotein vaccine. In: Lasky L, Fox CF (eds) Technological advances in vaccine development. Academic, New York, pp 455–469
Sanchez-Pescador L, Burke RL, Ott G, Van Nest G (1988b) The effect of adjuvants on the efficacy of a recombinant herpes simplex virus glycoprotein vaccine. J Immunol 141: 1720–1727
Schmersahl P, Rudiger G (1975) Behandlungsergebnisse mit dem Herpes simplex-Antigen Lupidon H bzw. Lupidon G. Z Hautarzt 50: 105
Skinner GRB, Woodman CBJ, Hartley CE, Buchan A, Fuller A, Durham J, Synnott M, Clay JC, Melling J, Wiblin C, Wilkins J (1982a) Preparation and immunogenicity of vaccine Ac NFU1 (S^-) MRC towards the prevention of herpes genitalis. Br J Venerol Dis 58: 381–386

Skinner GRB, Woodman CBJ, Hartley CE, Buchan A, Fuller A, Wilblin C, Wilkins G, Melling J (1982b) Early experience with "antigenoid" vaccine Ac Nfu_1, (S^-) MRC towards prevention or modification of herpes genitalis. Dev Biol Stand 52: 333–344

Skinner GRB, Fink CG, Cowan M, Buchan A, Fuller A, Hartley CE, Durham J, Wiblin C, Melling J (1987) Follow-up report on 50 subjects vaccinated against herpes genitalis with Skinner vaccine. Med Microbiol Immunol (Berl) 176: 161–168

Skinner GRB, Fink CG, Durham J et al. (1989) Follow-up report on 101 subjects vaccinated with Skinner herpes vaccine. In: Meheus A, Spier RE (eds) Vaccines for sexually transmitted disease. Butterworths, London, pp 202–207

Snapper SB, Lugosi L, Jekkel A, Melton RE, Kieser T, Bloom BR, Jacobs WR Jr (1988) Lysogeny and transformation in mycobacteria: stable expression of foreign genes. Proc Natl Acad USA 85: 6987–6991

Spear PG (1984) Glycoproteins specified by herpes simplex viruses. In: Roizman B (ed) Herpesviruses. Plenum, New York

Stanberry LR (1990) Subunit viral vaccines: prophylactic and therapeutic use. In: Aurelian L (ed) Herpesviruses, the immune system and aids. Kluwer Academic, Norwell, pp 309–341

Stanberry LR (1991) Herpes simplex virus vaccines. In: Feigin RD, Glezen WP (eds) Update on immunization: progress in vaccine development. Semin Pediatr Infect Dis 2: 178–185

Stanberry LR, Bernstein DI, Burke RL, Pachl C, Myers MG (1987) Vaccination with recombinant herpes simplex virus glycoproteins: protection against initial and recurrent genital herpes. J Infect Dis 155: 914–920

Stanberry LR, Burke RL, Meyers MG (1988) Herpes simplex virus glycoprotein treatment of recurrent genital disease. J Infect Dis 157: 156–163

Stanberry LR, Harrison CJ, Bernstein DI, Burke RL, Shukla R, Ott, G, Myers MG (1989) Herpes simplex virus glycoprotein immunotherapy of recurrent genital herpes: factors influencng efficacy. Antiviral Res 11: 203–214

Straus SE, Takiff HE, Seidlin M, Bachrach S, Lininger L, DiGiovanna JJ, Wester KA, Smith HA, Lehrman SN, Creagh-Kirk T, Alling DW (1984) Suppression of frequently recurring genital herpes. A placebo controlled double-blind trial of oral acyclovir. N Engl J Med 310: 1545–1550

Sullivan V, Smith G (1987) Expression and characterization of herpes simplex virus type 1 (HSV-1) glycoprotein G (gG) by recombinant vaccinia virus: neutralization of HSV-1 infectivity with anti-gG antibody. J Gen Virol 68: 2587–2598

Thompson RL, Nakashizuka M, Stevens JG (1986) Vaccine potential of a live avirulent herpes simplex virus. Microb Pathog 1: 409–416

Tigges MA, Koelle D, Hartog K, Sekulovich RE, Corey L, Burke RL (1992) Human $CD8^+$ herpes simplex virus-specific cytotoxic T lymphocyte clones recognize diverse virion protein antigens. J Virol (in press)

Valenzuela P, Coit D, Medina-Selby MA, Kuo CH, Van Nest G, Burke RL, Bull P, Urdea MS, Graves PV (1985) Antigen engineering in yeast: synthesis and assembly of hybrid hepatitis B surface antigen-herpes simplex 1 gD particles. Bio/Technology 3: 323–326

Wachsman M, Aurehian L, Smith CC, Lipinskas BR, Perkus ME, Poaletti E (1987) Protection of guinea pigs from primary and recurrent herpes simplex virus (HSV) type 2 cutaneous disease with vaccinia virus recombinants expressing HSV glycoprotein D. J Infect Dis 155(6): 1188–1197

Watson RJ, Enquist LW (1985) Genetically engineered herpes simplex virus vaccines. Prog Med Virol 31: 84–108

Watson RJ, Weis JH, Salstrom JS, Enquist LW (1982) Herpes simplex virus, type 1-glycoprotein D gene. Nucleotide sequence and expression in *Escherichia coli*. Science 218: 381–384

Weir JP, Bennett M, Allen EM, Elkins KL, Martin S, Rouse BT (1989) Recombinant vaccinia virus expressing the herpes simplex virus type 1 glycoprotein C protects mice against herpes simplex virus challenge. J Gen Virol 70: 2587–2594

Woodman CBJ, Buchan A, Fuller A, Hartley C, Skinner GRB, Stocker D, Sugrue D, Clay JC, Wilkins G, Wiblin C, Melling J (1983) Efficacy of vaccine Ac Nfu_1, (S^-) MRC given after an initial clinical episode in the prevention of herpes genitalis. Br J Venerol Dis 59: 311–313

Zarling JM, Moran PA, Brewer L, Ashley R, Corey L (1988) Herpes simplex virus (HSV)-specific proliferative and cytotoxic T-cell responses in humans immunized with an HSV type 2 glycoprotein subunit vaccine. J Virol 62: 4481–4485

Zweerink HJ, Martinez D, Lynch RJ, Stanton LW (1981) Immune responses in mice against herpes simplex virus: mechanisms of protection against facial and ganglionic infections. Infect Immun 31: 267–275

Controlling Herpes Simplex Virus Infections: is Intracellular Immunization the Way of the Future?

K. K. WONG, Jr.[1] and S. CHATTERJEE[2]

1 Introduction

The available approaches to the management of herpes simplex virus (HSV) infections currently consist primarily of chemotherapy and the possibility of preventive vaccination. In this review, we explore the potential of a third approach based on recent advances in molecular biology and genetic engineering which renders cells resistant to viral infection. This form of antiviral gene therapy, termed "intracellular immunization" by BALTIMORE (1988), involves the intracellular expression of a variety of molecular species specifically designed to inhibit targeted virus replication. The utility of such an approach lies in the direct application of knowledge pertaining to molecular mechanisms involved in viral pathogenesis to precisely interrupt lytic infection through the generation of virus-resistant

[1] Department of Hematology Bone Marrow Transplantation Service City of Hope National Medical Center Building W4 Room 404A 1500 East Duarte Road, Duarte, CA 91010-0269, USA
[2] Division of Pediatrics, City of Hope National Medical Center and Department of Molecular Genetics, Beckman Research Institute, 1500 East Duarte Road, Duarte, CA 91010-0269, USA

Current Topics in Microbiology and Immunology, Vol. 179

cells. Such strategies may be particularly useful in the management of intracellular pathogens which are capable of evading the immune system or in circumstances in which knowledge of molecular pathogenic mechanisms far surpasses the ability to develop appropriate therapeutic drugs or vaccines. In this article, the basic concepts of intracellular immunization, relevant aspects of HSV molecular biology, and studies utilizing antiviral gene therapy to prevent HSV infection will be reviewed.

2 Intracellular Immunization: Basic Concepts

Effective antiviral gene therapy is predicated upon a number of basic tenets which include: (1) the stable expression of gene(s) coding for molecular species produced in sufficient quantities to inhibit essential targeted viral functions, (2) the absence of cellular toxicity of the viral inhibitory species, and (3) the requirement for a highly efficient, nontoxic, gene delivery system. The chosen strategy would preferably be capable of conferring viral resistance to self-renewing stem cells, thus providing a perpetually resistant cell population. Although there is currently no consensus concerning the most effective target, inhibition of early viral functions, at a time when viral gene products are least abundant and perhaps most easily inhibited, may be desirable, thereby preventing the accumulation of late viral gene products which may be inimical to the cell. Additionally, inhibition of highly conserved essential gene functions may afford cellular protection from serotypically diverse but genetically related viruses. Thus, both HSV-1 and HSV-2 production are significantly inhibited in cells constitutively expressing an antisense RNA complementary to upstream sequences of the HSV-1 ICP4 transcript (Wong et al. 1991, see below). Fortunately, mutational analyses have identified a plethora of requisite virus-encoded regulatory proteins expressed early in viral replicative cycles which may constitute suitable targets, including the tat and rev proteins of human immunodeficiency virus (HIV), the tax protein of human T cell leukemia virus (HTLV-1), the E1A protein of adenovirus, the large T-antigen of SV40, and the ICP4 and ICP27 proteins of HSV (see below).

3 Viral Inhibitory Molecules and Examples of Successful Intracellular Immunization Against Other Viruses

3.1 General Approaches

Advances in intracellular immunization have closely paralleled our understanding of complex molecular, cellular, and viral processes. Thus, a variety of viral inhibitory molecules have been examined for efficacy in antiviral gene therapy, including: (1) truncated sense and/or antisense RNAs targeting integral viral mRNAs, (2) antisense-targeted catalytic RNAs (ribozymes) which cleave viral transcripts in a site-specific manner, (3) negative transdominant peptides or proteins that inhibit essential regulatory proteins, (4) viral receptor analogues or glycoproteins that interfere with viral attachment, cell penetration, or egress, and (5) interferon (IFN) or other antiviral molecules. These classes of viral inhibitory molecules, their proposed mechanisms of action, and examples of their use in the successful establishment of intracellular resistance against other viruses will be discussed briefly.

3.2 Antisense RNA

Naturally occurring antisense transcripts perform important regulatory roles in prokaryotes (for review see GREEN et al. 1986; SIMONS and KLECKNER 1988), have been developed as tools to specifically inhibit gene expression in eukaryotic cells (WEINTRAUB et al. 1985; VAN DER KROL et al. 1988a), produce phage resistance in bacteria (as a paradigm of an "artificial immune system") (COLEMAN et al. 1985), and inhibit pathogenic viral replication in eukaryotic cells in vitro (ZAMECNIK and STEPHENSON 1978; SMITH et al. 1986; GOODCHILD et al. 1988; KULKA et al. 1989) and transgenic plants in vivo (POWELL et al. 1989). Although the exact mechanism of inhibition of targeted gene expression is currently unknown, endogenously produced antisense RNA probably functions in a combinatorial fashion via formation of sense: antisense hybrids that: (1) are subject to accelerated degradation by intracellular nucleases, (2) inhibit nuclear to cytoplasmic mRNA transport, (3) potentially disrupt regulatory proteins: mRNA interactions, (4) inhibit cytoplasmic mRNA translation, (5) are subject to an intracellular unwinding function that introduces mutations (adenosine to inosine conversions), (6) inhibit posttranscriptional mRNA processing (i.e., splicing or polyadenylation) via specific antisense transcripts targeting these regions, and (7) may induce IFN expression (for reviews see GREEN et al. 1986; VAN DER KROL et al. 1988b; WALDER 1988; HELENE and TOULME 1990). In addition, a protein that unwinds RNA hybrids and in so doing, disrupts gene expression by mutating adenine to inosine residues has also recently been described in *Xenopus* oocytes and several types of mammalian cells (BASS and WEINTRAUB 1988; WALDER 1988). Intracellular expression of antisense transcripts complementary to essential viral transcripts has been effective in inhibiting Rous sarcoma virus (RSV) (CHANG and STOLTZFUS 1985, 1987), Moloney murine leukemia virus (MuLV) (SULLENGER et al. 1990a), human adenovirus type 5 (Ad5) (MIROSHNICHENKO et al. 1989), SV40 (JENNINGS and MOLLOY 1987), HTLV-1 (VON RUDEN and GILBOA 1989), and HIV-1 production (RHODES and JAMES 1990; SCZAKIEL and PAWLITA 1991; CHATTERJEE et al. 1991). However, antisense RNA has not been universally effective in abrogating targeted gene expression, perhaps due to several factors, including variability of antisense expression resulting from either low promoter strength or host chromosomal "position effects" flanking the inserted gene (ALLEN et al. 1988; VAN DER

KROL et al. 1988a), inhibition of antisense:sense hybrid formation due to steric interference from RNA secondary structure, and extreme potency of the gene product such that residual quantities possess sufficient effector function for phenotypic gene expression (reviewed in VAN DER KROL et al. 1988b).

3.3 Sense RNA

In contrast to the mechanisms proposed for antisense RNA-mediated inhibition of gene expression, intracellular expression of large quantities of truncated sense transcripts, containing essential, *cis*-active, regulatory regions, competes with wild-type mRNA for necessary, often rate-limiting, regulatory proteins, thereby inhibiting targeted gene expression and virus production. A requisite stem-loop sequence termed TAR (*trans*-activating response element) has been identified at the beginning of all HIV transcripts and has been demonstrated to be essential in mediating the regulatory role of tat, a virus-encoded protein that significantly augments HIV gene expression (FENG and HOLLAND 1988). Thus, intracellular synthesis of HIV-1 TAR-containing transcripts has been demonstrated to interrupt requisite tat-TAR interactions with subsequent inhibition of HIV long terminal repeat (LTR)-directed gene expression and virus production (GRAHAM and MAIO 1990; SULLENGER et al. 1990b; LISZIEWICZ et al. 1991). However, efficient abrogation of targeted gene expression utilizing this strategy may depend upon particularly abundant expression of individual or concatemeric sense transcripts.

3.4 Ribozymes

Ribozymes comprise a class of RNA molecules that were originally discovered through their unique autocatalytic (self-cleaving) activity (for reviews see CECH and BASS 1986; CECH 1988). Hammerhead ribozymes, originally described from a class of plant pathogens (PROUDY et al. 1986), have subsequently been modified for the sequence-specific cleavage of RNA in *trans* (intermolecular cleavage) (HASELOFF and GERLACH 1988). It should be noted that theoretically any RNA transcript containing a ribozyme cleavage sequence (GUX in which X can be C, A, or U) can be cut a hammerhead ribozyme targeted to the appropriate region by flanking complementary sequences. Thus, intracellular expression of genetically engineered hammerhead ribozymes has been utilized to cleave specific RNAs including the U7snRNA in frog oocytes (COTTON and BIRNSTIEL 1989) and an inserted chloramphenicol acetyl transferase (CAT) transcript, thereby abrogating its expression in monkey cells (CAMERON and JENNINGS 1989). Recently, the constitutive intracellular expression of an anti-HIV-1 gag transcript ribozyme inhibited HIV-1 antigen (p24 gag) production by >99% following viral challenge without manifesting associated cellular toxicity (SARVER et al. 1990). As ribozymes are targeted to the appropriate region of the transcript via flanking antisense sequences, they are subject to problems analogous to those previously described for antisense RNA-mediated inhibition of gene expression. Additional problems arise from the recognized pH (8–9) and temperature (50°–55 °C) optima of hammerhead ribozyme-mediated cleavage and the variance from conditions normally found within eukaryotic cells. However, the exciting potential of engineered ribozymes in the modulation of eukaryotic gene expression has prompted avid research into their modification so as to optimize cleavage under more physiologic conditions.

3.5 Transdominant Mutant Proteins

Inhibitory transdominant proteins capitalize upon the observation that regulatory proteins frequently consist of several, distinct, essential domains which are required for intermolecular interactions (i.e., with another protein or nucleic acids) (reviewed in PTASHNE 1988). Thus, transdominant mutants consisting exclusively of one domain can compete with the corresponding wild-type protein in the formation of functional multimeric complexes or for binding sites on DNA or RNA (HERSKOWITZ 1987). Utilizing this approach, mutant transdominant proteins have been developed which functionally inhibit several HIV regulatory proteins including tat (GREEN et al. 1989; PEARSON et al. 1990), rev (MALIM et al. 1989; MERMER et al. 1990; BENKO et al. 1990), gag (TRONO et al. 1989), the HTLV-1 rex protein (RIMSKY et al. 1989; BOHNLEIN et al. 1991), HSV VP16 (FRIEDMAN et al. 1988), and ICP4 (SHEPARD et al. 1990) (see below). Importantly, intracellular expression of several of these transdominants has been associated with significant protection from cytopathogenicity and inhibition of viral production following challenge with cognate virus (FRIEDMAN et al. 1988; TRONO et al. 1989; SHEPARD et al. 1990).

3.6 Viral Receptors and Receptor Analogs

Several approaches have involved the manipulation of viral receptors in the establishment of intracellular immunity. CD4 has been demonstrated to be the primary cellular receptor for HIV via its interaction with the viral envelope protein gp120 (DALGLEISH et al. 1984; KLATZMANN et al. 1984). MORGAN et al. (1990) demonstrated that cells transformed with a retroviral vector expressing soluble CD4 (sCD4) secreted sufficient quantities of sCD4 into the surrounding media to protect adjacent HIV-susceptible cells from a subsequent viral challenge. In contrast to the extracellular expression of sCD4, BUONOCORE and ROSE (1990) utilized a vaccinia vector to express CD4 to which the endoplasmic reticulum (ER) retention signal (KDEL) had been attached. Transiently expressed HIV gp120 complexed with ER-retained CD4, thereby preventing transit to the cell surface and significantly inhibiting syncytia formation. The theoretical utility of this novel approach stems from the utilization of the CD4 receptor as a common pathway for cellular entry of a serologically diverse group of pathogenic retroviruses, including both HIV-1 and HIV-2. However, actual inhibition of virus production following a viral challenge and the potential long-term cellular toxicity of large quantities of ER-retained proteins have not yet been determined.

3.7 IFN and Other Naturally Occurring Antiviral Molecules

Intracellular expression of IFNs and other naturally occurring antiviral molecules has also been studied. IFNs are a class of potent, pleiotropic, broad-spectrum, antiviral glycoproteins which are secreted by cells following a number of stimuli, including viral infection and exposure to double-stranded RNA. Although the exact mechanisms of antiviral action have yet to be elucidated, IFNs interact with specific receptors at the cell surface and induce the synthesis of several enzymes which appear to be important for the subsequent development of an "antiviral" state. One of these enzymes, a protein kinase termed P1, phosphorylates the α subunit of eukaryotic protein synthesis initiation factor-2 (eIF-2α) thereby inhibiting protein synthesis, while another, (2′,5′)-oligoadenylate synthetase, induces an endonuclease that degrades single-stranded RNA (for review see STAEHELI 1990). CD4-positive human lymphocytes transformed with a retroviral vector containing the IFN-α gene under control of the inducible HIV LTR expressed elevated levels of IFN-α following HIV challenge and restricted virus production by greater than 95% relative to control cells (BEDNARIK et al. 1989). A similar approach was utilized by SEIF et al. (1991) who transfected murine cells with a plasmid conferring G418 resistance and containing the murine IFN-β gene under control of the murine H-2Kb major histocompatibility complex (MHC) promoter. Clonally derived cells expressing IFN-β displayed enhanced resistance to vesicular stomatitis, encephalomyocarditis, and Semliki forest viruses in comparison to control cells. Augmented IFN expression offers the advantages of a broad antiviral spectrum associated with the ability to confer protection to nontransformed neighboring cells. However, the pleiotropic effects of IFNs, including their broad effects upon the immune system and cellular proliferation, associated with the recognized in vivo toxicity of continuous high concentrations of IFN (reviewed in GRESSER 1986), would necessitate tightly regulated expression and might hamper this interesting application in vivo.

In contrast, intracellular expression of a specific antigen-binding domain derived from a monoclonal antibody has been utilized to inhibit p21 ras protein expression in vitro (WERGE et al. 1990). A similar approach has been suggested by FARAJI-SHADAN et al. (1990) as a potential form of gene therapy directed against HIV and could prove useful in the intracellular inhibition of any antigen against which a monoclonal antibody could be raised. Although this approach, which perhaps provides "intracellular immunization" in the strictest sense, is promising, the molecular manipulations involved are quite complex, and actual inhibition of targeted virus production employing this unique strategy has not yet been demonstrated.

Finally, the results of ARNHEITER et al. (1990), who developed influenza-resistant transgenic mice via pronuclear transfer of the naturally occurring *Mx-1* resistance gene, and SALTER and CRITTENDEN (1989), who developed avian leukosis virus (ALV)-resistant transgenic chickens expressing the ALV envelope glycoprotein, are notable for demonstrating that the principles of intracellular immunization can be highly effective in vivo.

4 Gene Delivery Systems: Eukaryotic Viral Vectors

The efficacy of intracellular immunization is dependent upon both the overall effectiveness of the viral inhibitory species and the efficiency of the delivery system. Although a number of techniques exist for cellular delivery of DNA molecules, including transfection (see SAMBROOK et al. 1989 and references therein), electroporation (ANDREASON and EVANS 1988), and use of modified liposomes (MANNINO and GOULD-FOGERITE 1988), the high efficiency required for transforming the large number of cells necessary for effective antiviral gene therapy essentially precludes all but transduction with eukaryotic viral vectors. Consequently, vectors, useful for gene therapy are generally based upon noncytopathic eukaryotic viruses in which nonessential or readily complementable genes have been replaced with the gene of interest. Currently, vectors based upon noncytopathic retroviruses, whose life cycle involves reverse transcription of genomic viral RNA into DNA with subsequent proviral integration into host cellular DNA, have gained wide usage for genetic transfer and have recently been approved for human gene therapy trials (for reviews see ANDERSON 1984; EGLITIS and ANDERSON 1988; MILLER 1990). Indeed, retroviral vectors have recently been utilized to confer intracellular immunity in vitro against HTLV-1 (VON RUDEN and GILBOA 1989) and HIV-1 (RHODES and JAMES 1990; SULLENGER et al. 1990b) and have been proposed for the expression of CD4 for the inhibition of HIV-1 infection (MORGAN et al. 1990). While retroviruses do not appear to integrate into host DNA in a totally random fashion, proviral integration occurs in a relatively large number of sites and has been associated with both insertional mutagenesis (HARTUNG et al. 1986) and tumorigenesis due to insertion of viral transcriptional enhancer elements in proximity to endogenous cellular oncogenes (CUYPERS et al. 1984; HAYWARD et al. 1981). In addition, successful proviral integration is strongly dependent upon transduction (infection) of actively cycling cells, which may represent only a small fraction of the target cell population (MILLER et al. 1990).

Recently, vectors based upon the nonpathogenic defective parvovirus, adeno-associated virus (AAV), have been proposed for antiviral gene therapy (CHATTERJEE et al. 1991; WONG et al. 1991). AAV-based vectors offer several advantages such as: (1) lack of pathogenicity, (2) extensive host range, including a wide variety of species and tissues, (3) heat and lipid solvent stability, (4) high transduction (infection) frequencies, and (5) lack of superinfection inhibition thus allowing multiple series of transductions (HERMONAT and MUZYCZKA 1984; TRATSCHIN et al. 1984; LEBKOWSKI et al. 1988; LEFACE et al. 1988; MCLAUGHLIN et al. 1988; for review of AAV biology see CUKOR et al. 1984; BERNS and BOHENZKY 1987). Recent reports indicate that AAV DNA may integrate into human cellular DNA in a site-specific manner (KOTIN and BERNS 1989; KOTIN et al. 1990). This latter feature offers the advantage of minimizing the possibility of insertional mutagenesis and variability of inserted gene expression due to environmental influences arising from the surrounding host chromosome ("position effects") (ALLEN et al. 1988; VAN DER KROL 1988a). In addition, stable, latent, wild-type AAV infections have been followed in tissue culture for greater than 100 passages in the absence of selective pressure, implying that AAV genomic integration is a relatively stable event (BERNS et al. 1975). Finally, wild-type AAV has never been demonstrated to cause disease, an important feature when contemplating its use as a vector for gene therapy. However, whether AAV-based vectors offer significant advantages over well established retroviral vectors for the purpose of gene delivery awaits the results of ongoing comparative studies.

5 In Vitro HSV Molecular Targets

Just as effective antiviral vaccine development necessitates a thorough knowledge of the immune system and its response to viral challenge, successful intracellular immunization demands a detailed understanding of the molecular pathogenesis and cellular responses to targeted viral infection. Several in vitro studies have explored the feasibility of molecularly interrupting essential aspects of the HSV replicative cycle as a prelude to actual intracellular immunization. Thus, in a manner analogous to the inhibition of RSV production (ZAMECNIK and STEPHENSON 1978), exogenous administration of specific synthetic oligonucleotides has been shown to protect against HSV challenge. SMITH et al. (1986), and KULKA et al. (1989) have developed oligonucleotide methylphosphonates complementary to the acceptor splice junctions of HSV-1 ICP22 and ICP47 transcripts, immediate early (IE) genes that, interestingly, have not been shown to be essential for productive viral infection (see above). Nevertheless, these antisense oligonucleotides inhibited viral production by up to 98% in a sequence-specific manner, with a greater effect upon HSV-1 than HSV-2; restricted splicing of ICP22 pre-mRNA; and reduced viral DNA synthesis by 75% and protein synthesis by 90% without affecting cellular protein synthesis. Mechanistically it is difficult to reconcile the discrepancy between previous studies, which demonstrated that ICP22 and ICP47 were dispensable for lytic infection (MAVROMARA-NAZOS et al. 1986; POST and ROIZMAN 1981; SEARS et al. 1985), and the profound inhibition of HSV production observed in these studies. It is possible that inhibition of expression from another, as yet unidentified, virus-encoded or cellular spliced transcript essential for HSV lytic infection might have occurred, although this is purely speculative.

Utilizing a different strategy, GAO et al. (1989) demonstrated that HSV-2 DNA polymerase was readily inhibited by phosphorothioate homo-oligodeoxynucleotides in vitro. Experiments comparing oligonucleotide base composition and chain length demonstrated that the 28 base phosphorothioate oligodeoxycytidine S-(dC)28 displayed the greatest degree of inhibition and was competitive with respect to the DNA template. In contrast, HSV-1 KOS and human cellular DNA polymerases (α, β, γ) were comparatively less sensitive (but not totally insensitive) to the inhibitory effects of S-(dC)28. Importantly, phosphorothioate analogue-treated cells restricted HSV-2 production by >90% in comparison to controls following viral challenge, without associated cellular toxicity (GAO et al. 1990).

Attempts have also been made to inhibit HSV-encoded enzymes using synthetic peptides designed to interrupt the assembly of functional multimeric complexes. For example, HSV DNA codes for the expression of ribonucleotide reductase (RR) a hetero-oligomeric enzyme which catalyzes the conversion of ribonucleotides to deoxyribonucleotide diphosphates and which is required for efficient viral DNA synthesis in resting cells (GOLDSTEIN and WELLER 1988). Synthetic oligopeptides corresponding to the COOH-terminal region of the smaller RR2 subunit inhibited enzyme activity in vitro in a reversible fashion via the inhibition of formation of the active holoenzyme (COHEN et al. 1986; DUTIA et al. 1986; MCCLEMENTS et al. 1988; PARADIS et al. 1988). However, pretreatment of cell monolayers with these peptides did not protect against HSV challenge, perhaps because of poor cellular penetration (DUTIA et al. 1986).

As noted above, the earliest, essential virus-encoded function following HSV entry into the cell involves the transactivation of IE genes by a transcription complex comprised of the virion-associated protein VP16 and several cellular factors. Thus, HAIGH et al. (1990) demonstrated that a synthetic peptide comprising VP16 amino acid residues 360–373 strongly inhibited requisite VP16: cellular transcription factor (TRF) complex formation without affecting independent TRF DNA binding. However, the ability of this peptide to either inhibit HSV IE gene expression or to prevent HSV infection has not yet been demonstrated.

6 Intracellular Immunization and HSV

In contrast to the extracellular administration of viral inhibitory molecules, intracellular expression offers several innate advantages. Exposure to ubiquitous extracellular nucleases and proteases is eliminated, and difficulties which can arise from problems with cellular transport, as observed with the peptide which inhibited formation of the active holoenzyme of HSV ribonucleotide reductase (DUTIA et al. 1986), are obviated. However, relatively few studies have described intracellular expression of viral inhibitory molecules with attendant cellular protection from an HSV challenge.

JOHNSON and SPEAR (1989) inserted the HSV-1 KOS glycoprotein D (*gD1*) gene under control of the human metallothienein II promoter within a plasmid conferring resistance to the neomycin analogue geneticin (G418). Plasmid transfected, G418- selected, murine (LMtk-) and human (Hep-2) cells expressing gD1 were protected from HSV-1 KOS cytopathogenicity and restricted HSV-1 KOS production up to 99.9% in comparison to control cells following viral challenge (multiplicity of infection 0.25). Viral protein expression and HSV-mediated shutoff of cellular protein synthesis were reduced in HSV-1 KOS-challenged, gD1-expressing cells in comparison to controls. The level of protection correlated with the amount of gD1 expression and was serotype-specific; cytopathic effects following challenge with another strain of HSV-1 (HSV-1 F) and two strains of HSV-2 (HSV-2 G and 333) were not readily inhibited. Radiolabeled HSV-1 KOS virions attached equally well to all cell lines but failed to penetrate gD1-expressing cells, as evidenced by the absence of characteristic virus-mediated shutoff of host protein synthesis. Thus, the block of virus production was localized to the level of cell penetration and may have resulted from cell surface gD1 complexing with a molecule which was requisite for HSV penetration. Finally, gD1 expression did not appear to be inimical to cells in tissue culture. This demonstrated block of HSV penetration in cells expressing surface gD1 was proposed to have arisen as a mechanism to prevent multiple infections (superinfection) of the same cell, which would result in eclipse of progeny virus and a reduction in overall infection efficiencies.

To target the earliest, requisite, viral-encoded, regulatory function following cellular entry, FRIEDMAN et al. (1988) constructed a mutant gene by truncating the COOH-terminal 78 amino acids from the HSV-1 IE gene transactivator VP16 (or Vmw65) and inserted it within a plasmid under murine sarcoma virus (MSV) promoter control. This plasmid was cotransfected with another plasmid containing the HSV-1 *tk* gene into murine tk-deficient (tk-) L cells, and the cells were selected in hypoxanthine-aminopterin-thymidine (HAT) medium. Cells expressing this VP16 transdominant mutant protein displayed markedly diminished cytopathic effects and a 40-fold (98%) reduction in virus production in comparison to control cells following an HSV-1 challenge (multiplicity of infection 0.1). In addition, the observed inhibitory effect was specific for HSV-1, as pseudorabies virus, another herpes virus whose replicative cycle does not involve an IE gene transactivator analogous to VP16, replicated equally well in all cell lines. Additionally, expression of the mutant VP16 did not adversely affect cell viabilities, potentially paving the way for in vivo studies.

As noted above, ICP4 expression occurs soon after viral entry and is required throughout the replicative cycle for productive HSV infection. In contrast to the use of transfection techniques, WONG and CHATTERJEE (WONG et al. 1991) developed a eukaryotic viral vector based upon adeno-associated virus (see above) which conferred G418 resistance and was designed to express an antisense transcript complementary to a portion of the 5′-noncoding leader and first coding AUG of the ICP4 transcript. Vector-transduced (infected), clonal, murine L929 cells expressing this transcript were protected from HSV-1 cytopathogenicity, restricted HSV-1 production by 1000- to 10000-fold, and exhibited prolonged survival (75%–80% viability by trypan blue dye exclusions vs 0% by day 4 postinfection) in comparison to control cells following viral challenge (multiplicity of infection 0.1). However, vaccinia virus, a genetically distinct virus, replicated to an equivalent extent in all cell lines. ICP4 antisense expression did not affect cell proliferation or viability, and the level of protection was stable for a period of greater than 6 months. Interestingly, HSV-2 production from HSV-1 ICP4 antisense-expressing cells was inhibited to a similar extent following viral challenge, reflecting the high level of nucleic acid sequence conservation (about 80%) between the two viruses in the region targeted by the antisense transcript.

Finally, SHEPARD et al. (1990) constructed a plasmid conferring G418 resistance and expressing a mutant HSV *ICP4* gene product (termed X25) under inducible ICP4 promoter control. The mutant ICP4 product, which lacked amino acids 1–32 and 275–774, formed heterodimers with the wild-type protein, bound to DNA, and autoregulated *ICP4* gene expression, but was devoid of transactivation properties. Furthermore, plasmid-transfected G418-selected cells expressing this mutant protein displayed a 94% reduction in HSV-1 production in comparison to controls following viral challenge (multiplicity of infection 0.3). The degree of protection was multiplicity-dependent, was more pronounced against HSV than pseudorabies virus (although it should be emphasized that the expression of the transdominant ICP4 molecule was under inducible control and that pseudorabies virus does not contain an IE gene transactivator analogous to VP16), and probably resulted from a combination of interference with wild-type ICP4 DNA binding and inhibition of functional ICP4 heterodimer formation.

7 Naturally Occurring HSV Resistance

SARMIENTO and KLEINERMAN (1990) have recently described herpesvirus-specific induction of 57 kDa and 62 kDa proteins within human peripheral blood monocytes and lymphocytes which may play a role in conferring natural resistance to HSV. These proteins were identified in cells which were naturally highly resistant to HSV infection but not in lymphocytes which were either permissive or partially permissive. The block in viral replication occurred early in the replicative cycle, prior to the onset of HSV IE gene expression. However, the exact role of these proteins in the development of viral resistance has yet to be fully delineated, while efforts are underway to isolate and further characterize the genes which express them. If these genes prove to be important in the establishment of innate HSV resistance, their expression, either individually or in combination to determine whether viral resistance can be conferred to an otherwise susceptible cell population, in a fashion analogous to the influenza virus/*Mx-1* gene paradigm of mice (ARNHEITER et al. 1990; STAEHELI et al. 1986) would be of tremendous interest.

8 Perspectives and Future Considerations

Much can be learned from the studies of intracellular immunization directed against other viruses. A wide variety of viral inhibitory molecules have been expressed either constitutively or inducibly without associated cellular toxicity. Thus, the intracellular expression of antisense-targeted ribozymes or truncated sense RNAs targeting either essential HSV transcripts, antigen recognition regions of monoclonal antibodies raised against essential gene products, or viral receptors and receptor analogues may prove useful in antiviral gene therapy directed against HSV and related viruses. In addition, exogenous administration of IFN has been highly effective in inhibiting HSV production, probably at a stage following viral entry and prior to IE gene expression (DOMKE et al. 1986; MITTNACHT et al. 1988). Thus, insertion of a IFN gene under control of an early HSV-inducible promoter, such as an IE gene promoter, in a manner analogous to the studies of BEDNARIK et al. (1989) and SEIF et al. (1991) described above, may also be highly effective against HSV. As previously summarized, the complex molecular biology of HSV replication has been studied in some detail and a wide variety of potential targets for antiviral gene therapy has been delineated. Accordingly, intracellular expression of molecular species specifically designed to disrupt VP16 and/or IE gene expression or function, ribonucleotide reductase, virally encoded genes necessary for DNA synthesis (DNA polymerase, UL9, UL42, UL5, 8, 52, ICP8), assembly of infectious particles, and reactivation from the latent state will continue to remain of great interest, particularly following efficacy determinations in vivo.

No discussion of intracellular immunization or gene therapy would be complete without mention of potential ethical problems and safety issues. The actual genetic manipulation of somatic and germ cells in vitro and in vivo carries with it a great responsibility to ensure the safety of all involved. Thus, the use of multiply defective, nonpathogenic, viral vectors to eliminate potential horizontal spread, utilization of appropriate containment facilities, and strict adherence to well-conceived regulations similar to the NIH guidelines for recombinant DNA research cannot be overemphasized.

In contrast to the development of an effective smallpox vaccine, which will shortly celebrate its bicentennial anniversary (JENNER 1798; BREMAAN and ARITA 1980), intracellular immunization is a relatively young field; most studies have been described only within the last 3–5 years. The choice of an optimal gene delivery system, use of constitutive or inducible promoters, optimization of viral inhibitory species either individually or in combination, tissue-specific targeting or expression, frequency of emergence of viral "escape" mutants, the stability and specificity of protection, and the potential for toxicity are vital issues which need to be resolved. Would transduction of neural cells with a vector designed to express a molecule which strongly inhibits a requisite early HSV function in vitro, such as ICP4 or ICP27, protect against subsequent viral challenge and/or the development of a latent state? Similarly, would a vector designed to abolish ICP0 expression disrupt the transition from latent to productive HSV infection, thereby inducing a "permanent", clinically silent latent state? Studies designed to address these issues and determine the efficacy and potential toxicity of intracellular immunization in vivo, via utilization of viral vectors and animal models or the generation of transgenics, are currently underway. However, although many questions remain as yet unanswered, the basic foundation for effective intracellular immunization has been laid. Just as Edward Jenner did almost 200 years ago, we now stand at the threshold of an exciting new era in the control of viral diseases that have plagued humanity for thousands of years. Only the passage of time will reveal whether intracellular immunization will be equally successful.

References

Allen ND, Cran DG, Barton SC, Hettle S, Reik R, Surani MAH (1988) Transgenes as probes for active chromosomal domains in mouse development. Nature 333: 852–855
Anderson WF (1984) Prospects for human gene therapy. Science 226: 401–409
Andreason GL, Evans GA (1988) Introduction and expression of DNA molecules in eukaryotic cells by electroporation. BioTechniques 6: 650–660
Anheiter H, Skuntz S, Noteborn M, Chang S, Meler E (1990) Transgenic mice with intracellular immunity to influenza virus. Cell 62: 51–61
Baltimore D (1988) Intracellular immunization. Nature 335: 395–396
Bass BL, Weintraub H (1988) An unwinding activity that covalently modifies its double-stranded RNA substrate. Cell 55: 1089–1098
Bednarik DP, Mosca JD, Raj NBK, Pitha PM (1989) Inhibition of human immunodeficiency virus (HIV) replication by HIV-trans-activated alpha-2 interferon. Proc Natl Sci USA 86: 4958–4962
Benko DM, Robinson R, Solomin L, Mellini M, Felber BK, Pavlakis GN (1990) Binding of trans-dominant mutant rev protein of human immunodeficiency virus type 1 to the cis-acting rev-responsive element does not affect the fate of viral mRNA. New Biologist 2: 1111–1122
Berns KI, Bohenzky RA (1987) Adeno-associated viruses: an update. Adv Virus Res 32: 243–306
Berns KI, Pinkerton TC, Thomas GF, Hoggan MD (1975) Detection of adeno-associated virus (AAV)-specific nucleotide sequences in DNA isolated from latently infected Detroit 6 cells. Virology 68: 556–560
Bohnlein S, Pirker FP, Hofer L, Zimmermann K, Bachmayer H, Bohnlein E, Hauber J (1991) Trans-dominant repressors for human T-cell leukemia virus type I rex and human immunodeficiency virus type 1 rev function. J Virol 65: 81–88
Breman JG, Arita I (1980) The confirmation and maintenance of smallpox eradication. N Engl J Med 303: 1263–1273
Buonocore L, Rose JK (1990) Prevention of HIV-1 glycoprotein transport by soluble CD4 retained in the endoplasmic reticulum. Nature 345: 625–628
Cameron FH, Jennings PA (1989) Specific gene suppression by engineered ribozymes in monkey cells. Proc Natl Acad Sci USA 86: 9139–9143
Cech TR (1988) Ribozymes and their medical implications. J Am Med Assoc 260: 3030–3034
Cech TR, Bass BL (1986) Biological catalysis by RNA. Annu Rev Biochem 55: 599–629
Chang L-J, Stoltzfus CM (1985) Gene expression from both intronless and intron-containing Rous sarcoma virus clones is specifically inhibited by anti-sense RNA. Mol Cell Biol 5: 2341–2348
Chang L-J, Stoltzfus CM (1987) Inhibition of Rous sarcoma virus replication by antisense RNA. J Virol 61: 921–924
Chatterjee S, Rose JA, Johnson PR, Wong KK Jr (1991) Transduction of intracellular resistance to HIV production by an adeno-associated virus-based antisense vector. Vaccine 91: 85–90
Cohen EA, Gaudreau P, Brazeau P, Langelier Y (1986) Specific inhibition of herpesvirus ribonucleotide reductase by a nonapeptide derived from the carboxy terminus of subunit 2. Nature 321: 441–443
Coleman J, Hirashima A, Inokuchi Y, Green PJ, Inouye M (1985) A novel immune system against bacteriophage infection using complementary RNA (micRNA). Nature 315: 601–603
Cotton M, Birnstiel ML (1989) Ribozyme mediated destruction of RNA in vivo. EMBO J 8: 3861–3866
Cukor G, Blacklow NR, Hoggan D, Berns KI (1984) Biology of adeno-associated virus. In: Berns KI (ed) The Parvoviruses. Plenum, New York, pp 33–66
Cuypers HT, Selten G, Quint W, Zijlstra M, Maandag ER, Boelens W, van Wezenbeek P, Melief C, Berns A (1984) Murine leukemia virus-induced T-cell lymphomagensis: integration of proviruses in a distinct chromosomal region. Cell 37: 141–150

Dalgleish AG, Beverley PCl, Clapham PR, Crawford DH, Greaves MF, Weiss RA (1984) The CD4 (T4) antigen is an essential component of the receptor for the AIDS retrovirus. Nature 312: 763–767
Domke I, Straub P, Kirchner H (1986) Effect of interferon on replication of herpes simplex virus type 1 and 2 in human macrophages. J Virol 60: 37–42
Dutia BM, Frame MC, Subak-Sharpe JH, Clark WN, Marsden HS (1986) Specific inhibition of herpesvirus ribonucleotide reductase by synthetic peptides. Nature 321: 439–441
Eglitis MA, Anderson WF (1988) Retroviral vectors for introduction of genes into mammalian cells. BioTechniques 6: 608–614
Faraji-Shadan F, Stubbs JD, Bowman PD (1990) A putative approach for gene therapy against human immunodeficiency virus (HIV). Med Hypotheses 32: 81–84
Feng S, Holland EC (1988) HIV-1 tat trans-activation requires the loop sequence within tar. Nature 334: 165–167
Friedman AD, Triezenberg SJ, McKnight SL (1988) Expression of a truncated viral trans-activator selectively impedes lytic infection by its cognate virus. Nature 335: 452–454
Gao W-Y, Stein CA, Cohen JS, Dutschman GE, Cheng Y-C (1989) Effect of phosphorothioate homo-oligodeoxynucleotides on herpes simplex virus type 2 induced DNA polymerase. J Biol Chem 264: 11521–11526
Gao W-Y, Hanes RN, Vazquez-Padua MA, Stein CA, Cohen JS, Cheng Y-C (1990) Inhibition of herpes simplex virus type 2 growth by phosphorothioate oligodeoxynucleotides. Antimicrob Agents Chemother 34: 808–812
Goldstein DJ, Weller SK (1988) Factor(s) present in herpes simplex virus type 1-infected cells can compensate for the loss of the large subunit of the viral ribonucleotide reductase: characterization of an ICP6 deletion mutant. Virology 166: 41–51
Goodchild J, Agrawal S, Civiera MP, Sarin PS, Sun D, Zamecnik PC (1988) Inhibition of human immunodeficiency virus replication by antisense oligodeoxynucleotides. Proc Natl Acad Sci USA 85: 5507–5511
Graham GJ, Maio JJ (1990) RNA transcripts of the human immunodeficiency virus transactivation response element can inhibit action of the viral transactivator. Proc. Natl Acad Sci USA 87: 5817–5821
Green M, Ishino M, Loewenstein PM (1989) Mutational analysis of HIV-1 tat minimal domain peptides: identification of trans-dominant mutants that suppress HIV-LTR-driven gene expression. Cell 58: 215–223
Green PJ, Pines O, Inouye M (1986) The role of antisense RNA in gene regulation. Annu Rev Biochem 55: 569–97
Gresser I (1986) Interferon-induced diseases. In: Notkins AL, Oldstone MBA (eds) Concepts in viral pathogenesis II. Springer, Berlin Heidelberg New York, pp 232–242
Haigh A, Greaves R, O'Hare P (1990) Interference with the assembly of a virus-host transcription complex by peptide competition. Nature 334: 257–259
Hartung S, Jaenisch R, Breindl M (1986) Retrovirus insertion inactivates mouse al(I) collagen gene by blocking initiation of transcription. Nature 320: 365–367
Haseloff J, Gerlach WL (1988) Simple RNA enzymes with new and highly specific endoribonuclease activities. Nature 334: 585–591
Hayward WS, Neel BG, Astrin SM (1981) Activation of a cellular onc gene by promoter insertion in ALV-induced lymphoid leukosis. Nature 290: 475–480
Helene C, Toulme J-J (1990) Specific regulation of gene expression by antisense, sense and antigene nucleic acids. Biochim Biophys Acta 1049: 99–125
Hermonat PL, Muzyczka N (1984) Use of adeno-associated virus as a mammalian DNA cloning vector: transduction of neomycin resistance into mammalian tissue culture cells. Proc Natl Acad Sci USA 81: 6466–6470
Herskowitz I (1987) Functional inactivation of genes by dominant negative mutations. Nature 329: 219–222
Jenner E (1798) An inquiry into the causes and effects of the variolae vaccine, a disease discovered in some of the western counties of England, particularly Gloucestershire, and known by the name of cowpox. Sampson Low, London
Jennings PD, Molloy PL (1987) Inhibition of SV40 replicon function by engineered antisense RNA transcribed by RNA polymerase III. EMBO J 10: 3043–3047
Johnson RM, Spear PG (1989) Herpes simplex virus glycoprotein D mediates interference with herpes simplex virus infection. J Virol 63: 819–827
Klatzmann D, Champagne E, Chamaret S, Gruest J, Guetard D, Hercend T, Gluckman J-C, Montagnier L (1984) T-lymphocyte T4 molecule behaves as the receptor for human retrovirus LAV. Nature 312: 767–768

Kotin RM, Berns KI (1989) Organization of adeno-associated virus DNA in latently infected Detroit 6 cells. Virology 170: 460–467

Kotin RM, Siniscalco M, Samulski RJ, Zhu X, Hunter L, Laughlin CA, McLaughlin S, Muzyczka N, Rocchi M, Berns KI (1990) Site-specific integration by adeno-associated virus. Proc Natl Acad Sci USA 87: 2211–2215

Kulka M, Smith CC, Aurelian L, Fishelevich R, Meade K, Miller P, Ts'o POP (1989) Site specificity of the inhibitory effects of oligo (nucleoside methylphosphonate)s complementary to the acceptor splice junction of herpes simplex virus type 1 immediate early mRNA 4. Proc Natl Acad Sci USA 86: 6868–6872

LaFace D, Hermonat P, Wakeland E, Peck A (1988) Gene transfer into hematopoietic progenitor cells mediated by an adeno-associated virus vector. Virology 162: 483–486

Lebkowski JS, McNally MM, Okarma TB, Lerch LB (1988) Adeno-associated virus: a vector system for efficient introduction and integration of DNA into a variety of mammalian cell types. Mol Cell Biol 8: 3988–3996

Lisziewicz J, Rappaport J, Dhar R (1991) Tat-regulated production of multimerized TAR RNA inhibits HIV-1 gene expression. New Biologist 3: 82–89

Malim MH, Bohnlein S, Hauber J, Cullen BR (1989) Functional dissection of the HIV-1 rev trans-activator: derivation of a trans-dominant repressor of rev function. Cell 58: 205–214

Mavromara-Nazos P, Ackermann M, Roizman B (1986) Construction and properties of viable herpes simplex virus type 1 recombinant lacking coding sequences of the α47 gene. J Virol 60: 807–812

Mermer B, Felber BK, Campbell M, Pavlakis GN (1990) Identification of trans-dominant HIV-1 rev protein mutants by direct transfer of bacterially produced proteins into human cells. Nucleic Acids Res 18: 2037–2044

Mannino RJ, Gould-Fogerite S (1988) Liposome mediated gene transfer. BioTechniques 6: 682–690

McClements W, Yamanaka G, Garsky V, Perry H, Bacchetti S, Colonno R, Stein RB (1988) Oligopeptides inhibit the ribonucleotide reductase of herpes simplex virus by causing subunit separation. Virology 162: 270–273

McLaughlin SK, Collis P, Hermonat PL, Muzyczka N (1988) Adeno-associated virus general transduction vectors: analysis of proviral structures. J Virol 62: 1963–1973

Miller AD (1990) Progress toward human gene therapy. Blood 76: 271–278

Miller DG, Adam MA, Miller AD (1990) Gene transfer by retrovirus vectors occurs only in cells that are actively replicating at the time of infection. Mol Cell Biol 10: 4239–4242

Miroshnichenko OI, Ponomareva TI, Tikchonenko TI (1989) Inhibition of adenovirus 5 replication in COS-1 cells by antisense RNAs against the viral Ela region. Gene 84: 83–89

Mittnacht S, Straub P, Kirschner H, Jacobsen H (1988) Interferon treatment inhibits onset of herpes simplex virus immediate-early transcription. Virology 164: 201–210

Morgan RA, Looney DJ, Muenchau DD, Wong-Staal F, Gallo RC, Anderson WF (1990) Retroviral vectors expressing soluble CD4: a potential gene therapy for AIDS. AIDS Res Hum Retroviruses 6: 183–191

Paradis H, Gaudreau P, Brazeau P, Langelier Y (1988) Mechanisms of inhibition of herpes simplex virus (HSV) ribonucleotide reductase by a nonapeptide corresponding to the carboxyl terminus of its subunit 2. J Biol Chem 263: 16045–16050

Pearson L, Garcia J, Wu F, Modesti N, Nelson J, Gaynor R (1990) A transdominant tat mutant that inhibits tat-induced gene expression from the human immunodeficiency virus long terminal repeat. Proc Natl Acad Sci USA 87: 5079–5083

Post LE, Roizman B (1981) A generalized technique for deletion of specific genes in large genomes: α gene 22 of herpes simplex virus 1 is not essential for growth. Cell 25: 227–232

Powell PA, Stark DM, Sanders PR, Beachy RN (1989) Protection against tobacco mosaic virus in transgenic plants that express tobacco mosaic virus antisense RNA. Proc Natl Acad Sci USA 86: 6949–6952

Prody GA, Bakos JT, Buzayan JM, Schneider IR, Bruening G (1986) Autolytic processing of dimeric plant virus satellite RNA. Science 231: 1577–1580

Ptashne M (1988) How eukaryotic transcriptional activators work. Nature 335: 683–689

Rhodes A, James W (1990) Inhibition of human immunodeficiency virus replication in cell culture by endogenously synthesized antisense RNA. J Gen Virol 71: 1965–1974

Rimsky L, Duc Dodon M, Dixon EP, Greene WC (1989) Trans-dominant inactivation of HTLV-I and HIV-1 gene expression by mutation of the HTLV-I rex transactivator. Nature 341: 453–456

Salter DW, Crittenden LB (1989) Artificial insertion of a dominant gene for resistance to avian leukosis virus into the germ line of the chicken. Theor Appl Genet 77: 457–461

Sambrook J, Fritsch EF, Maniatis T (1989) Molecular cloning, a laboratory manual, 2nd edn. Cold Spring Harbor Laboratory Press, Cold Spring Harbor

Sarmiento M, Kleinerman ES (1990) Innate resistance to herpes simplex virus infection. Human lymphocyte and monocyte inhibition of viral replication. J Immunol 144(5): 1942–1953

Sarver N, Cantin EM, Chang PS, Zaia JA, Ladne PA, Stephens DA, Rossi JJ (1990) Ribozymes as potential anti-HIV-1 therapeutic agents. Science 247: 1222–1225

Sczakiel G, Pawlita M (1991) Inhibition of human immunodeficiency virus type 1 replication in human T cells stably expressing antisense RNA. J Virol 65: 468–472

Sears AE, Halliburton IW, Meignier B, Silver S, Roizman B (1985) Herpes simplex virus 1 mutant deleted in the α22 gene: growth and gene expression in permissive and restrictive cells, and establishment of latency in mice. J Virol 55: 338–346

Seif I, de Maeyer E, Riviere I, de Maeyer-Guignard J (1991) Stable antiviral expression in BALB/c 3T3 cells carrying a beta interferon sequence behind a major histocompatibility complex promoter fragment. J Virol 65: 664–671

Shepard AA, Polentino P, DeLuca NA (1990) Trans-dominant inhibition of herpes simplex virus transcriptional regulatory protein ICP4 by heterodimer formation. J Virol 64: 3916–3926

Simons RW, Kleckner N (1988) Biological regulation by antisense RNA in prokaryotes. Annu Rev Genet 22: 567–601

Smith CC, Aurelian L, Reddy MP, Miller PS, Ts'o POP (1986) Antiviral effect of an oligo(nucleoside methylphosphonate) complementary to the splice junction of herpes simplex virus type 1 immediate early pre-mRNAs 4 and 5. Proc Natl Acad Sci 83: 2787–2791

Staeheli P (1990) Interferon-induced proteins and the antiviral state. Adv Virus Res 38: 147–200

Staeheli P, Haller O, Boll W, Lindenmann J, Weissmann C (1986) Mx protein: constitutive expression in 3T3 cells transformed with cloned Mx cDNA confers selective resistance to influenza virus. Cell 44: 147–158

Sullenger BA, Lee TC, Smith CA, Ungers GE, Gilboa E (1990a) Expression of chimeric tRNA-driven antisense transcripts renders NIH3T3 cells highly resistant to Moloney murine leukemia virus replication. Mol Cell Biol 10: 6512–6523

Sullenger BA, Gallardo HF, Ungers GE, Gilboa E (1990b) Overexpression of TAR sequences renders cells resistant to human immunodeficiency virus replication. Cell 63: 601–608

Tratschin JD, West MH, Sandbank T, Carter BJ (1984) A human parvovirus, adeno-associated virus, as a eukaryotic vector: transient expression and encapsidation of the prokaryotic gene for chloramphenicol acetyltransferase. Mol Cell Biol 4: 2072–2081

Trono D, Feinberg MB, Baltimore D (1989) HIV-1 gag mutants can dominantly interfere with the replication of the wild-type virus. Cell 59: 113–120

Van der Krol AR, Lenting PE, Veenstra J, Van der Meer IM, Koes RE, Gerats AGM, Mol JNM, Stuitje AR (1988a) An anti-sense chalcone synthase gene in transgenic plants inhibits flower pigmentation. Nature 333: 866–869

Van der Krol AR, Mol JNM, Stuitje AR (1988b) Modulation of eukaryotic gene expression by complementary RNA or DNA sequences. BioTechniques 6: 958–976

Von Ruden T, Gilboa E (1989) Inhibition of human T-cell leukemia virus type 1 replication in primary human T cells that express antisense RNA. J Virol 63: 677–682

Walder J (1988) Antisense DNA and RNA: progress and prospects. Genes Dev 2: 502–504

Weintraub H, Izant JG, Harland RM (1985) Anti-sense RNA as a molecular tool for genetic analysis. Trends Genet 2: 22–25

Werge TM, Biocca S, Cattaneo A (1990) Intracellular immunization. Cloning and intracellular expression of a monoclonal antibody to the p21ras protein. FEBS Lett 274: 193–198

Whitley RJ (1990) Herpes simplex viruses. In: Fields BN, Knipe DM et al. Virology, 2nd edn. Raven, New York, pp 1843–1887

Wong KK Jr, Rose JA, Chatterjee S (1991) Restriction of HSV-1 production in cell lines transduced with an antisense viral vector targeting the ICP4 gene. In: Vaccine 91. Cold Spring Harbor Laboratory Press, New York, pp 183–189

Zamecnik PC, Stephenson ML (1978) Inhibition of Rous sarcoma virus replication and cell transformation by a specific oligodeoxynucleotide. Proc Natl Acad Sci USA 75: 280–284

Subject Index

Current Topics in Microbiology and Immunology

Volumes published since 1987 (and still available)

Vol. 135: **Paige, Christopher J.; Gisler, Roland H. (Ed.):** Differentiation of B Lymphocytes. 1987. 25 figs. IX, 150 pp. ISBN 3-540-17470-2

Vol. 136: **Hobom, Gerd; Rott, Rudolf (Ed.):** The Molecular Biology of Bacterial Virus Systems. 1988. 20 figs. VII, 90 pp. ISBN 3-540-18513-5

Vol. 137: **Mock, Beverly; Potter, Michael (Ed.):** Genetics of Immunological Diseases. 1988. 88 figs. XI, 335 pp. ISBN 3-540-19253-0

Vol. 138: **Goebel, Werner (Ed.):** Intracellular Bacteria. 1988. 18 figs. IX, 179 pp. ISBN 3-540-50001-4

Vol. 139: **Clarke, Adrienne E.; Wilson, Ian A. (Ed.):** Carbohydrate-Protein Interaction. 1988. 35 figs. IX, 152 pp. ISBN 3-540-19378-2

Vol. 140: **Podack, Eckhard R. (Ed.):** Cytotoxic Effector Mechanisms. 1989. 24 figs. VIII, 126 pp. ISBN 3-540-50057-X

Vol. 141: **Potter, Michael; Melchers, Fritz (Ed.):** Mechanisms in B-Cell Neoplasia 1988. Workshop at the National Cancer Institute, National Institutes of Health, Bethesda, MD, USA, March 23–25, 1988. 1988. 122 figs. XIV, 340 pp. ISBN 3-540-50212-2

Vol. 142: **Schüpach, Jörg:** Human Retrovirology. Facts and Concepts. 1989. 24 figs. 115 pp. ISBN 3-540-50455-9

Vol. 143: **Haase, Ashley T.; Oldstone Michael B. A. (Ed.):** In Situ Hybridization 1989. 22 figs. XII, 90 pp. ISBN 3-540-50761-2

Vol. 144: **Knippers, Rolf; Levine, A. J. (Ed.):** Transforming. Proteins of DNA Tumor Viruses. 1989. 85 figs. XIV, 300 pp. ISBN 3-540-50909-7

Vol. 145: **Oldstone, Michael B. A. (Ed.):** Molecular Mimicry. Cross-Reactivity between Microbes and Host Proteins as a Cause of Autoimmunity. 1989. 28 figs. VII, 141 pp. ISBN 3-540-50929-1

Vol. 146: **Mestecky, Jiri; McGhee, Jerry (Ed.):** New Strategies for Oral Immunization. International Symposium at the University of Alabama at Birmingham and Molecular Engineering Associates, Inc. Birmingham, AL, USA, March 21–22, 1988. 1989. 22 figs. IX, 237 pp. ISBN 3-540-50841-4

Vol. 147: **Vogt, Peter K. (Ed.):** Oncogenes. Selected Reviews. 1989. 8 figs. VII, 172 pp. ISBN 3-540-51050-8

Vol. 148: **Vogt, Peter K. (Ed.):** Oncogenes and Retroviruses. Selected Reviews. 1989. XII, 134 pp. ISBN 3-540-51051-6

Vol. 149: **Shen-Ong, Grace L. C.; Potter, Michael; Copeland, Neal G. (Ed.):** Mechanisms in Myeloid Tumorigenesis. Workshop at the National Cancer Institute, National Institutes of Health, Bethesda, MD, USA, March 22, 1988. 1989. 42 figs. X, 172 pp. ISBN 3-540-50968-2

Vol. 150: **Jann, Klaus; Jann, Barbara (Ed.):** Bacterial Capsules. 1989. 33 figs. XII, 176 pp. ISBN 3-540-51049-4

Vol. 151: **Jann, Klaus; Jann, Barbara (Ed.):** Bacterial Adhesins. 1990. 23 figs. XII, 192 pp. ISBN 3-540-51052-4

Vol. 152: **Bosma, Melvin J.; Phillips, Robert A.; Schuler, Walter (Ed.):** The Scid Mouse. Characterization and Potential Uses. EMBO Workshop held at the Basel Institute for Immunology, Basel, Switzerland, February 20–22, 1989. 1989. 72 figs. XII, 263 pp. ISBN 3-540-51512-7

Vol. 153: **Lambris, John D. (Ed.):** The Third Component of Complement. Chemistry and Biology. 1989. 38 figs. X, 251 pp. ISBN 3-540-51513-5

Vol. 154: **McDougall, James K. (Ed.):** Cytomegaloviruses. 1990. 58 figs. IX, 286 pp. ISBN 3-540-51514-3

Vol. 155: **Kaufmann, Stefan H. E. (Ed.):** T-Cell Paradigms in Parasitic and Bacterial Infections. 1990. 24 figs. IX, 162 pp. ISBN 3-540-51515-1

Vol. 156: **Dyrberg, Thomas (Ed.):** The Role of Viruses and the Immune System in Diabetes Mellitus. 1990. 15 figs. XI, 142 pp. ISBN 3-540-51918-1

Vol. 157: **Swanstrom, Ronald; Vogt, Peter K. (Ed.):** Retroviruses. Strategies of Replication. 1990. 40 figs. XII, 260 pp. ISBN 3-540-51895-9

Vol. 158: **Muzyczka, Nicholas (Ed.):** Viral Expression Vectors. 1992. 20 figs. IX, 176 pp. ISBN 3-540-52431-2

Vol. 159: **Gray, David; Sprent, Jonathan (Ed.):** Immunological Memory. 1990. 38 figs. XII, 156 pp. ISBN 3-540-51921-1

Vol. 160: **Oldstone, Michael B. A.; Koprowski, Hilary (Ed.):** Retrovirus Infections of the Nervous System. 1990. 16 figs. XII, 176 pp. ISBN 3-540-51939-4

Vol. 161: **Racaniello, Vincent R. (Ed.):** Picornaviruses. 1990. 12 figs. X, 194 pp. ISBN 3-540-52429-0

Vol. 162: **Roy, Polly; Gorman, Barry M. (Ed.):** Bluetongue Viruses. 1990. 37 figs. X, 200 pp. ISBN 3-540-51922-X

Vol. 163: **Turner, Peter C.; Moyer, Richard W. (Ed.):** Poxviruses. 1990. 23 figs. X, 210 pp. ISBN 3-540-52430-4

Vol. 164: **Bækkeskov, Steinnun; Hansen, Bruno (Ed.):** Human Diabetes. 1990. 9 figs. X, 198 pp. ISBN 3-540-52652-8

Vol. 165: **Bothwell, Mark (Ed.):** Neuronal Growth Factors. 1991. 14 figs. IX, 173 pp. ISBN 3-540-52654-4

Vol. 166: **Potter, Michael; Melchers, Fritz (Ed.):** Mechanisms in B-Cell Neoplasia 1990. 143 figs. XIX, 380 pp. ISBN 3-540-52886-5

Vol. 167: **Kaufmann, Stefan H. E. (Ed.):** Heat Shock Proteins and Immune Response. 1991. 18 figs. IX, 214 pp. ISBN 3-540-52857-1

Vol. 168: **Mason, William S.; Seeger, Christoph (Ed.):** Hepadnaviruses. Molecular Biology and Pathogenesis. 1991. 21 figs. X, 206 pp. ISBN 3-540-53060-6

Vol. 169: **Kolakofsky, Daniel (Ed.):** Bunyaviridae. 1991. 34 figs. X, 256 pp. ISBN 3-540-53061-4

Vol. 170: **Compans, Richard W. (Ed.):** Protein Traffic in Eukaryotic Cells. Selected Reviews. 1991. 14 figs. X, 186 pp. ISBN 3-540-53631-0

Vol. 171: **Kung, Hsing-Jien; Vogt, Peter K. (Eds.):** Retroviral Insertion and Oncogene Activation. 1991. 18 figs. X, 179 pp. ISBN 3-540-53857-7

Vol. 172: **Chesebro, Bruce W. (Ed.):** Transmissible Spongiform Encephalopathies. 1991. 48 figs. X, 288 pp. ISBN 3-540-53883-6

Vol. 173: **Pfeffer, Klaus; Heeg, Klaus; Wagner, Hermann; Riethmüller, Gert (Ed.):** Function and Specificity of γ/δ T Cells. 1991. 41 figs. XII, 296 pp. ISBN 3-540-53781-3

Vol. 174: **Fleischer, Bernhard; Sjögren, Hans Olov (Eds.):** Superantigens. 1991. 13 figs. IX, 137 pp. ISBN 3-540-54205-1

Vol. 175: **Aktories, Klaus (Ed.):** ADP-Ribosylating Toxins. 1992. 23 figs. IX, 148 pp. ISBN 3-540-54598-0

Vol. 176: **Holland, John J. (Ed.):** Genetic Diversity of RNA Viruses. 1992. 34 figs. IX, 226 pp. ISBN 3-540-54652-9

Vol. 177: **Müller-Sieburg, Christa; Torok-Storb, Beverly; Visser, Jan; Storb, Rainer (Eds.):** Hematopoietic Stem Cells. 1992. 18 figs. XIII, 143 pp. ISBN 3-540-54531-X

Vol. 178: **Parker, Charles J. (Ed.):** Membrane Defenses Against Attack by Complement and Perforins. 1992. 26 figs. VIII, 188 pp. ISBN 3-540-54653-7

MIX
Papier aus verantwortungsvollen Quellen
Paper from responsible sources
FSC® C105338

If you have any concerns about our products,
you can contact us on
ProductSafety@springernature.com

In case Publisher is established outside the EU,
the EU authorized representative is:
Springer Nature Customer Service Center GmbH
Europaplatz 3, 69115 Heidelberg, Germany

Printed by Libri Plureos GmbH
in Hamburg, Germany